HEINEMANN BIOLOGY 2

SKILLS AND ASSESSMENT

Yvonne Sanders

VCE Units 3 & 4

Written to the VCE Biology Study Design 2022–2026

Pearson Australia
(a division of Pearson Australia Group Pty Ltd)
459–471 Church Street, Level 1, Building B,
Richmond, Victoria 3121
PO Box 23360, Melbourne, Victoria 8012
www.pearson.com.au

First published 2021 by Pearson Australia
2027 2026 2025 2024
1 2 3 4 5 27 26 25 24

Project Leads: Misal Belvedere and Malcolm Parsons
Content Developer: Rebecca Wood
Lead Development Editors: Fiona Cooke and Zoe Hamilton
Project Managers: Shaun Birtles and Michelle Thomas
Production Editors: Natalie Lincoln and Virginia O'Brien
Editor: Fiona Maplestone
Cover Designer: Anne Donald
Designer: Anne Donald
Proofreader: Laura Rentsch
Production Services Design Analyst: Jennifer Johnston
Rights and Permissions Editors: Siân Human and Madeleine Roberts
Illustrators: Bruce Rankin, DiacriTech, Anne Donald
Printed in Australia by Pegasus Media and Logistics

ISBN 978 0 6557 0026 5

Pearson Australia Group Pty Ltd ABN 40 004 245 943

Disclaimer
The selection of internet addresses (URLs) provided for this book was valid at the time of publication and was chosen as being appropriate for use as a secondary education research tool. However, due to the dynamic nature of the internet, some addresses may have changed, may have ceased to exist since publication, or may inadvertently link to sites with content that could be considered offensive or inappropriate. While the authors and publisher regret any inconvenience this may cause readers, no responsibility for any such changes or unforeseeable errors can be accepted by either the authors or the publisher.

Indigenous Australians
Some of the images used in the *Heinemann Biology 2 Skills and Assessment* book might have associations with deceased Indigenous Australians. Please be aware that these images might cause sadness or distress in Aboriginal or Torres Strait Islander communities.

Practical activities
All practical activities, including the illustrations, are provided as a guide only and the accuracy of such information cannot be guaranteed. Teachers must assess the appropriateness of an activity and take into account the experience of their students and facilities available. Additionally, all practical activities should be trialled before they are attempted with students and a risk assessment must be completed. All care should be taken and appropriate personal protective clothing and equipment should be worn when carrying out any practical activity. Although all practical activities have been written with safety in mind, Pearson Australia and the authors do not accept any responsibility for the information contained in or relating to the practical activities, and are not liable for loss and/or injury arising from or sustained as a result of conducting any of the practical activities described in this book.

Attributions
The following abbreviations are used in this list: t = top, b = bottom, l = left, r = right, c = centre.

123RF: Eric Isselee, pp. 156tr, 21tl. AAP: David Crosling, p. 175bl; Ove Hoegh-Guldberg, University of QLD, p. 155br; Tom Sweeney, Star Tribune, p. 23tr. Alamy Stock Photo: Suzy Bennett, p. 188tl; Alexander Kondakov, p. 104cl; Ivan Kuzmin, p. 175cl; Ingo Oeland, p. 175b; Bjorn Svensson, p. 154tr; Universal Images Group North America LLC, pp. 145tr, 148t, 162tr; Bosiljka Zutich, p. 154br. Bugwood.org: Whitney Cranshaw, Colorado State University, p. 105tl. Fotolia: Susan Flashman, p. 156br. Genome Research Limited: Genome Research Limited, p. 44b. Getty Images: Cultura RM Exclusive/GIPhotoStock/Cultura Exclusive, p. 188tr. National Institute on Drug Abuse: 'How Does Drug Abuse Effect the HIV Epidemic?' Retrieved from www.drugabuse.gov/publications/research-reports/hivaids/how-does-drug-abuse-affect-hiv-epidemic, June 2020, p. 132b. Pearson Education Ltd: Oxford Designers & Illustrators Ltd, p. 155tr. Science Photo Library (SPL): Imgo Arndt/Nature Picture Library, pp. 94, 140-141, 201; Dr Jeremy Burgess, p. 65tr; CNRI, pp. 1, 53; Nigel Downer, p. 1; Science VU/B. John, Visuals Unlimited, p. xxv. Shutterstock: A Aleksii, p. 69tr; Bildagentur Zoonar GmbH, p. 175tl; BlueRingMedia, p. 104tl; Crystal Eye Studio, p. 118cr; Craig Dingle, p. 175t; Ger Bosma Photos, p. 156b; Damian Herde, p. 104bl; magnetix, p. 8tr; Marcio Jose Bastos Silva, p. 191t; Soleil Nordic, p. 92b; Alexander Weickart, p. 25cr. State Government of Victoria: 'Flora & Fauna Guarantee Action Statement 1993', Department of Sustainability and Environment, no. 49, 2003, p. 195c. University of Ottawa: Illustrated by J. Soucie © BIODIDAC (used with permission), p. 90c.

Contents

Contents

ISBN 978 0 6557 0026 5

AREA OF STUDY 2

How are species related over time?

AREA OF STUDY 3

How is scientific inquiry used to investigate cellular processes and/or biological change?

ISBN 978 0 6557 0026 5

How to use this book

The *Heinemann Biology 2 Skills and Assessment* book provides the opportunity to practise, apply and extend your learning through a range of supportive and challenging activities. These activities reinforce key concepts and skills and enable a flexible approach to learning. There are also regular opportunities for reflection and self-evaluation in the final worksheet in each area of study.

This resource has been written to the VCE Biology Study Design 2022–2026 and is divided into five areas of study—two in Unit 3 and three in Unit 4.

The first four areas of study consist of four main sections:

- key knowledge
- worksheets
- practical activities
- past VCE exam questions.

Area of Study 3 in Unit 4 supports development of the key science skills that you need to successfully design and conduct a scientific investigation.

BIOLOGY TOOLKIT

The Biology toolkit supports development of the skills and techniques required to undertake primary- and secondary-sourced investigations, and covers examination techniques and study skills. It also includes checklists, models, exemplars and scaffolded steps. The toolkit can serve as a reference tool to be consulted as needed.

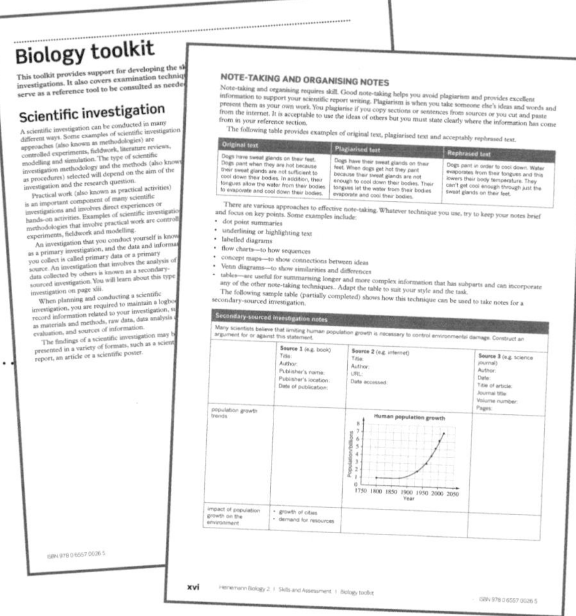
Biology toolkit

Scientific investigation

NOTE-TAKING AND ORGANISING NOTES

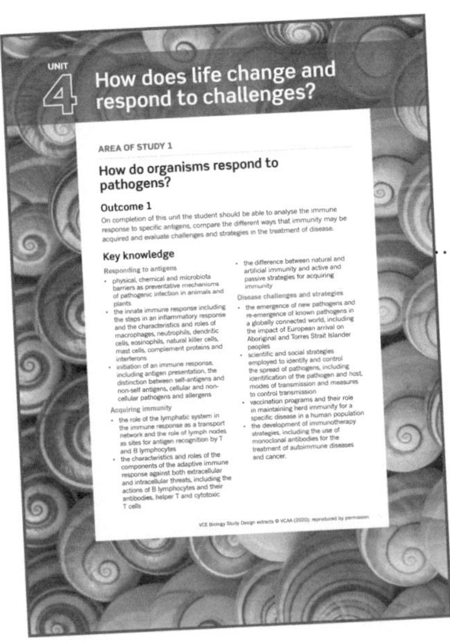
UNIT 4 How does life change and respond to challenges?

AREA OF STUDY 1

How do organisms respond to pathogens?

Outcome 1

Key knowledge

UNIT AND AREA OF STUDY OPENER

Heinemann Biology 2 Skills and Assessment is structured to follow the study design units and areas of study. The area of study opening page lists the key knowledge for easy reference to the activities that follow.

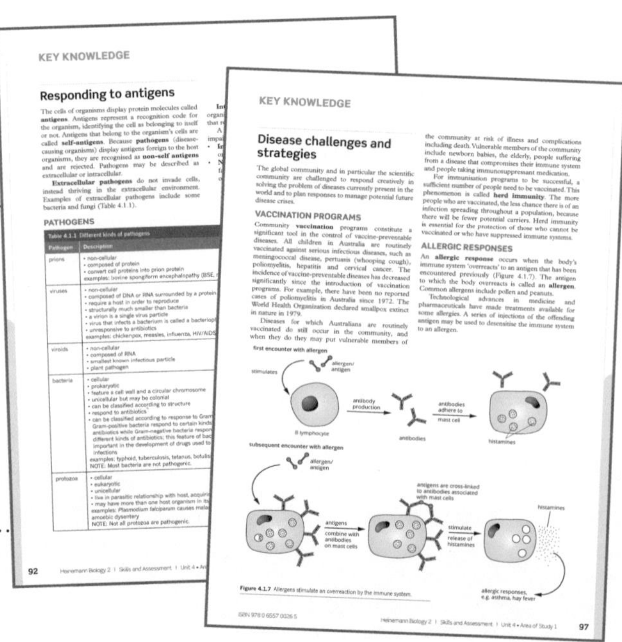
KEY KNOWLEDGE

Responding to antigens

KEY KNOWLEDGE

Disease challenges and strategies

KEY KNOWLEDGE

Each area of study begins with a key knowledge section. This consists of a set of summary notes that cover the key knowledge for that area of study. Key terms are in bold and are included in the glossary of the student book. The section also serves as a ready reference for completing the worksheets and practical activities.

ISBN 978 0 6557 0026 5

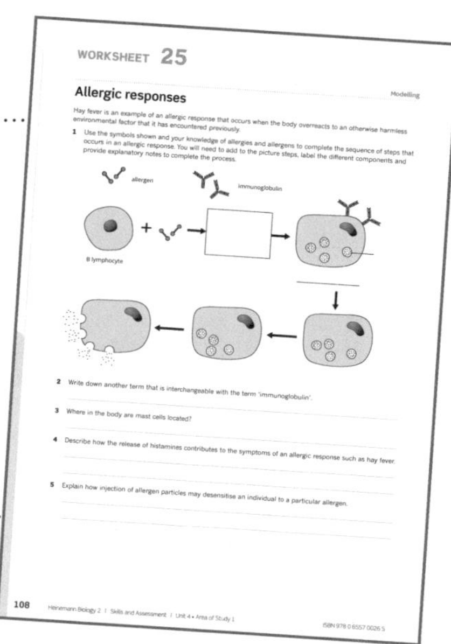
WORKSHEET 25

Modelling

Allergic responses

Hay fever is an example of an allergic response that occurs when the body overreacts to an otherwise harmless environmental factor that it has encountered previously.

1 Use the symbols shown and your knowledge of allergies and allergens to complete the sequence of steps that occurs in an allergic response. You will need to add to the picture steps, label the different components and provide explanatory notes to complete the process.

2 Write down another term that is interchangeable with the term 'immunoglobulin'.

3 Where in the body are mast cells located?

4 Describe how the release of histamines contributes to the symptoms of an allergic response such as hay fever.

5 Explain how injection of allergen particles may desensitise an individual to a particular allergen.

108 Heinemann Biology 2 | Skills and Assessment | Unit 4 • Area of Study 1 ISBN 978 0 6557 0026 5

WORKSHEETS

The worksheets feature questions that allow you to practise and apply your knowledge and skills. Each area of study includes a 'Knowledge review' worksheet, to activate prior knowledge, and a 'Reflection' worksheet, which you can use for self-assessment. Other worksheets provide opportunities to revise, consolidate and further your understanding.

All worksheets function as formative assessment and are clearly aligned with the study design.

PRACTICAL ACTIVITY 9

Controlled experiment

Investigating the effect of antibiotics on bacterial growth

Suggested duration: 30 minutes on first lesson; 60 minutes on second lesson

INTRODUCTION

Mastrings or Multodiscs are commercially produced discs impregnated with a variety of different antibiotic substances. Antibiotics are chemicals that deter the growth of bacteria. Different kinds of bacteria are affected by different kinds of antibiotics. Each satellite on the Mastring contains a different antibiotic that is identified by either a colour or a letter code or both.

AIM

- To investigate the effect of antibiotics on bacterial growth.

METHOD

1 • Collect three sterile nutrient agar Petri dishes. Keep the lids in place and use a marker pen to label the three dishes A, B—*E. coli*, and C—*S. albus*. Use initials or some other code to identify your Petri dishes from others in the class. It is a good idea to label the underside of the dish only, and at the perimeter of the dish rather than the middle.

2 • Collect a container of each of the two bacteria and two sterile swabs. Your teacher will demonstrate how to prepare a 'lawn culture' of the bacteria. Use the swabs to prepare bacterial cultures of *E. coli* and *S. albus* in Petri dishes B and C respectively. Immediately replace the lids. Place swabs in a disposal bag.

3 • Collect two Mastrings. Use fine forceps to place one Mastring in the centre of Petri dish B, as shown in Figure 4.1.10. Replace the lid. Repeat this procedure for Petri dish C.

MATERIALS

- disposable gloves
- 3 sterile, prepared nutrient agar Petri dishes
- broth cultures of *Escherichia coli* (*E. coli*) and *Staphylococcus albus* (*S. albus*)
- 2 sterile swabs in container
- marker pen
- fine forceps
- clear adhesive tape or Parafilm M laboratory tape
- access to incubator set at 35°C
- 2 sterile Mastrings
- disposal bags
- paper towel
- disinfectant
- access to soap and water for washing hands

Figure 4.1.9 Bacterial culture showing colonies after incubation

Mastring

agar plate

Figure 4.1.10 Mastring on agar plate

4 • Seal each of the Petri dishes with four short pieces of clear adhesive tape, placed at 12 o'clock, 3 o'clock, 6 o'clock and 9 o'clock. Alternatively, seal the Petri dishes with a strip of Parafilm M laboratory tape.

5 • Set the Petri dishes in place to be incubated at 35°C for 24–36 hours or at room temperature for three days. This will depend on your class timetable and teacher discretion.

6 • When all equipment is put away, wipe down your bench using paper towel and disinfectant. Then wash your hands thoroughly.

112 Heinemann Biology 2 | Skills and Assessment | Unit 4 • Area of Study 1 ISBN 978 0 6557 0026 5

PRACTICAL ACTIVITIES

Practical activities offer you the chance to complete practical work related to the various themes covered in the study design. You have the opportunity to design and conduct scientific investigations, generate, evaluate and analyse data, appropriately record results and prepare evidence-based conclusions. Where relevant, you will also need to conduct risk assessments to identify any potential hazards.

Each practical activity includes a suggested duration. Together with the Area of Study 3 practical investigation, the practical activities meet the 30 hours of practical work mandated for Units 3 and 4 in the study design.

Controlled experiment

METHODOLOGIES

Each worksheet and practical activity is mapped to one or more of the scientific investigation methodologies outlined in the study design. Completing these activities gives you experience in applying the methodologies in a wide variety of contexts and prepares you for designing and conducting your own scientific investigation in Unit 4 Area of Study 3.

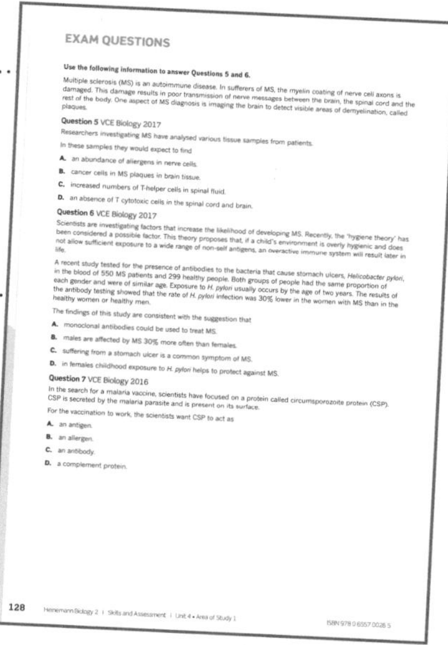
EXAM QUESTIONS

Use the following information to answer Questions 5 and 6.

Multiple sclerosis (MS) is an autoimmune disease. In sufferers of MS, the myelin coating of nerve cell axons is damaged. This damage results in poor transmission of nerve messages between the brain, the spinal cord and the rest of the body. One aspect of MS diagnosis is imaging the brain to detect visible areas of demyelination, called plaques.

Question 5 VCE Biology 2017

Researchers investigating MS have analysed various tissue samples from patients.

In these samples they would expect to find

A. an abundance of allergens in nerve cells.

B. cancer cells in MS plaques in brain tissue.

C. increased numbers of T-helper cells in spinal fluid.

D. an absence of T cytotoxic cells in the spinal cord and brain.

Question 6 VCE Biology 2017

Scientists are investigating factors that increase the likelihood of developing MS. Recently, the 'hygiene theory' has been considered a possible factor. This theory proposes that, if a child's environment is overly hygienic and does not allow sufficient exposure to a wide range of non-self antigens, an overactive immune system will result later in life.

A recent study tested for the presence of antibodies to the bacteria that cause stomach ulcers, *Helicobacter pylori*, in the blood of 550 MS patients and 299 healthy people. Both groups of people had the same proportion of each gender and were of similar age. Exposure to *H. pylori* usually occurs by the age of two years. The results of the antibody testing showed that the rate of *H. pylori* infection was 30% lower in the women with MS than in the healthy women or healthy men.

The findings of this study are consistent with the suggestion that

A. monoclonal antibodies could be used to treat MS.

B. males are affected by MS 30% more often than females.

C. suffering from a stomach ulcer is a common symptom of MS.

D. in females childhood exposure to *H. pylori* helps to protect against MS.

Question 7 VCE Biology 2016

In the search for a malaria vaccine, scientists have focused on a protein called circumsporozoite protein (CSP). CSP is secreted by the malaria parasite and is present on its surface.

For the vaccination to work, the scientists want CSP to act as

A. an antigen.

B. an allergen.

C. an antibody.

D. a complement protein.

128 Heinemann Biology 2 | Skills and Assessment | Unit 4 • Area of Study 1 ISBN 978 0 6557 0026 5

EXAM QUESTIONS

Each area of study finishes with a selection of past VCE Biology exam questions. This gives you the opportunity to gain valuable experience applying your knowledge and understanding to real exam questions.

TEACHER SUPPORT

Comprehensive answers and fully worked solutions for all worksheets, practical activities and exam questions are provided via the *Heinemann Biology 2 eBook* or Pearson Places. In-depth support for Unit 4 Area of Study 3 in the form of samples, templates and teacher notes is also included, along with an interactive SPARKlab for every practical activity.

Your one-stop shop for VCE Biology success

Heinemann Biology 6th edition, written to the new VCE Biology Study Design 2022–2026, offers a seamless experience for teachers and learners.

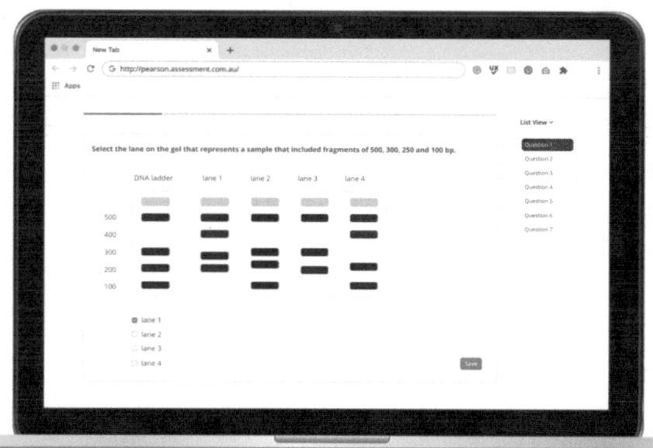

Enhanced eBook with assessment

Save time by accessing content and assessment in one simple-to-use platform.

Teacher access

- Digital assessment tool for informing planning, assigning work, tracking progress and identifying gaps
- Comprehensive sample teaching plan
- Suite of practice exam materials
- Rich support for area of study 3
- Solutions and support for the skills and assessment book

Learner access

- Videos and interactives
- Digital assessment questions with scaffolded hints, instant feedback and auto-correction
- Student book fully worked solutions
- SPARKlab versions of the skills and assessment book practical activities, developed in collaboration with PASCO

Student book

The **student book** addresses the latest developments and applications in biology. It uses best-practice literacy and learning design making the content and concepts accessible to all learners.

Key features include:

- Case studies with real-world data and analysis questions
- A smooth progression from low to high order questions in section, chapter and area of study reviews

Skills and assessment book

The **skills and assessment book** gives students the edge in applying key science skills and preparing for all forms of assessment.

Key features include:

- A skills toolkit
- Key knowledge study notes
- Worksheets
- Practical activities
- VCAA exam-style questions for each area of study
- Sample area of study 3 investigations

Find out more at **pearson.com.au/heinemannvce**

Biology toolkit

This toolkit provides support for developing the skills required to undertake scientific investigations. It also covers study skills and examination preparation. The toolkit can serve as a reference tool to be consulted as needed.

Scientific investigation

A scientific investigation can be conducted in many ways. Examples of scientific investigation approaches (also known as methodologies) are controlled experiments, literature reviews and modelling. The scientific investigation methodology and the methods (also known as procedures) selected will depend on the aim of the investigation and the research question.

Practical work (or practical activities) involves direct experiences or hands-on activities. Scientific investigation methodology involves all elements of planning—it considers the focus of the investigation and the rationale for approaching investigations in a particular way, for example, through controlled experiments, fieldwork or modelling. See the checklist on page x and page xiii for further information about methodology versus method in scientific investigations.

An investigation that you conduct yourself is known as a primary investigation, and the data and information you collect is called primary data or a primary source. An investigation that involves the analysis of data collected by others is known as a secondary-sourced investigation (see page xv).

When planning and conducting a scientific investigation, you must maintain a logbook to record information related to your investigation, such as materials and methods, raw data, data analysis, and sources of information.

The findings of a scientific investigation may be presented in a variety of formats, such as a scientific report, an article or a scientific poster.

CONDUCTING A SCIENTIFIC INVESTIGATION

Scientific investigations follow a precise scientific method. The checklist on the following page provides a summary of the elements that are common to many scientific investigation methodologies and scientific reports. Refer to the checklist and record important information as you conduct your scientific investigation.

Elements	Explanation	Tick ✔
date	the date the scientific investigation was conducted and the report was written	
title/heading	a statement of what is being investigated	
aim	• a statement that describes what the scientist wants to investigate, demonstrate or find out • can be written as 'to investigate the effect of *x* on *y*'	
hypothesis	• a tentative explanation for an observation that is based on evidence and prior knowledge • must be testable • written as a short statement, a hypothesis should include the variables to be tested and a statement of the measurable, predicted outcome • can be written as 'if *x* happens, then *y* will happen'. The 'if' part of the hypothesis relates to the independent variable. The 'then' part relates to the dependent variable.	
variables	• experiments measure relationships between variables • an independent variable is the variable that is changed • a dependent variable is the variable responding to the change in the independent variable • controlled variables are kept constant throughout the experiment	
risk assessment/ safety	• an analysis of the potential risks of conducting the investigation • objective is to maximise safety by managing and minimising risk (the risk assessment form on page xii guides you through the analysis of risk)	
materials	• a list of all equipment, chemicals and materials used • includes quantity, volume, concentration and mass of materials used	
methodology	a description of the overall approach undertaken in the scientific investigation and the reasons for taking this approach	
method	• a detailed, numbered, step-by-step list of exactly what is to be done • includes details about the materials used at each step (e.g. volume and concentration of a chemical) • may include scientific drawings or labelled diagrams • includes enough detail to allow the experiment to be repeated exactly • considers steps that can be taken to reduce errors	
results	• a record of all observations and measurements taken during the investigation • observations may be recorded as text, diagrams, photos or videos • data is commonly presented in tables and graphs and can include calculations • data should be checked for errors	
discussion	• interpretation of data, linking results to biological concepts • includes analysis of the data, looking for trends and relationships • identifies problems such as bias, inaccuracy, unreliable data and errors • evaluates the method used, including a discussion of limitations • comments on whether the results support or do not support the hypothesis • comments on the implications of the investigation and further investigations that may be undertaken	
conclusion	• a short statement based on evidence that summarises the findings of the investigation • provides a response to the research question or hypothesis • identifies the extent to which the investigation addressed the research question or hypothesis • no new information is introduced in the conclusion	

ISBN 978 0 6557 0026 5

RISK ASSESSMENT

Five levels of safety should be considered before carrying out a scientific investigation. The inverted pyramid ranks these levels in order of importance. The school and your teacher are responsible for reducing most of these risks. As a student scientist you can take measures to reduce the risks shown at the bottom of the hierarchy.

A risk assessment should be completed prior to conducting an investigation to determine the risks involved. Complete the risk assessment form on page xii to identify possible risks for which you can take responsibility, and to think of ways you can reduce risks to create a safe environment.

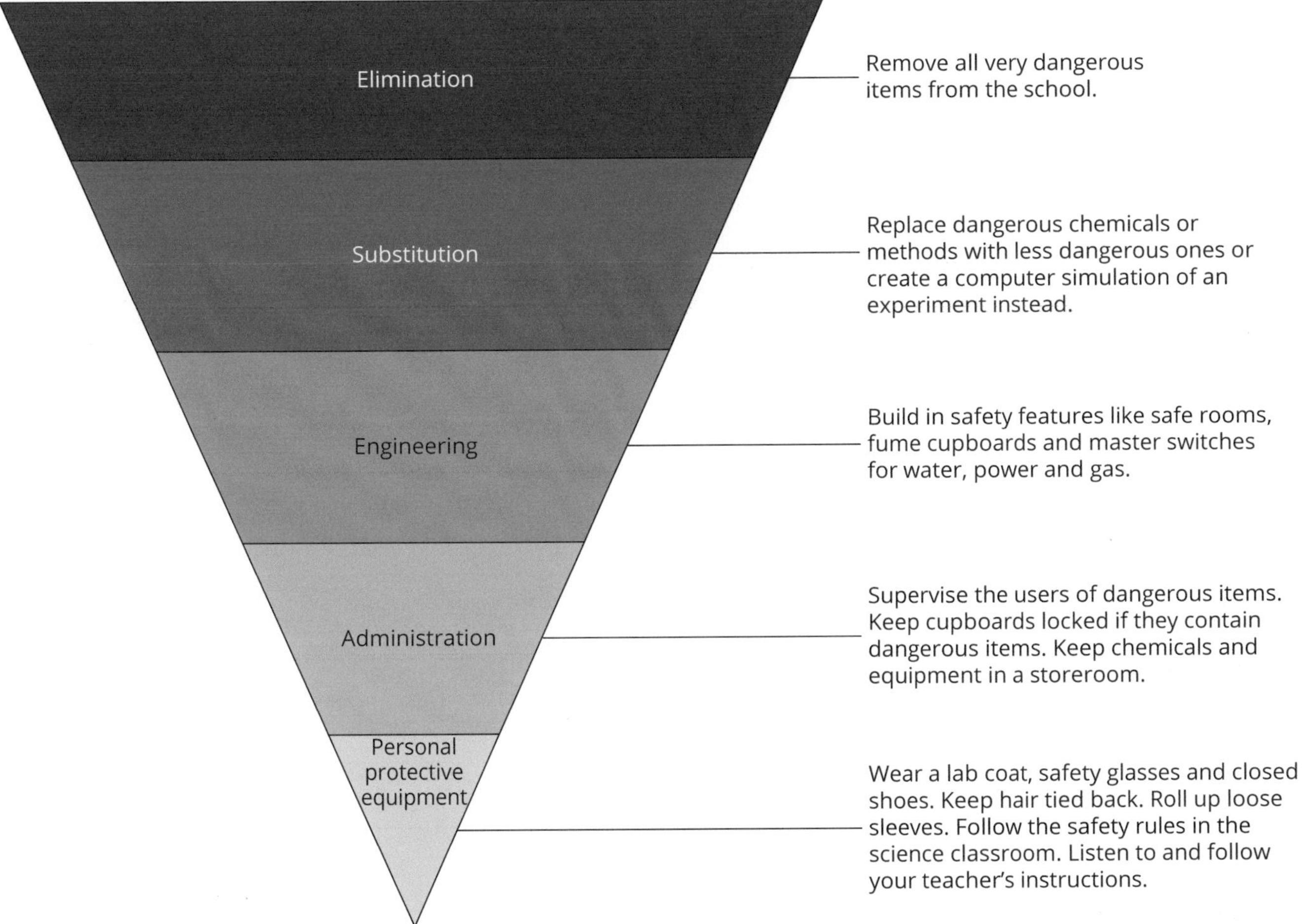

ISBN 978 0 6557 0026 5

Risk assessment form		
What activity are you doing?		
Title or description of the investigation:		
List ...	**Identify any risks**	**State how you will ...**
equipment you will be using:		use each piece:
chemicals you will be using:		carefully use each chemical: carefully dispose of the chemicals:
ethical issues you need to consider:		ethically use animals in the laboratory: ethically use human participants in the investigation:
outdoor or fieldwork activities:		reduce these risks:
any other possible risks:		reduce these risks:

ISBN 978 0 6557 0026 5

EXAMPLES OF SCIENTIFIC REPORTS

It can be difficult to gauge whether you have attained a high standard in your completed scientific report. Looking at sample scientific reports can help you identify what is required. Two sample scientific reports are provided in the *Heinemann Biology 1 Skills and Assessment* book. They include annotations to draw your attention to key points to note on each scientific report. These points are also reflected in the checklist, so you are able to use this as a tool to evaluate whether all requirements of the scientific investigation are complete.

STUDENT-DESIGNED INVESTIGATION

You will be required to design and conduct a scientific investigation based on the concepts you have learnt in either Unit 3 or Unit 4, or across both Units 3 and 4. This assessment task gives you the opportunity to apply key science skills and to pursue an area of interest to you, based on the key knowledge addressed during the course. You will be required to develop a question that drives your investigation, state an aim and hypothesis, select appropriate methodology and methods, and generate and collect primary quantitative data, recording important information in your logbook. You will then present your investigation as a scientific poster with a maximum word count of 600 words. It will be helpful to refer to page x to review what to include in a formal scientific report.

It will be important to carefully select the appropriate methodology and methods for your investigation. You will need to be clear about the difference between the two.

- Methodology: describes the overall approach undertaken in a scientific investigation; it considers the investigation more broadly, and includes the reasons for taking the chosen approach, for example, it will identify and describe strategies, such as the methods used to obtain data and the reasons why this is important to achieve the aim of your scientific investigation and address the question under investigation.
- Method: the specific procedure or steps taken to collect data during a scientific investigation.

DETERMINING YOUR RESEARCH QUESTION

As you develop a question for scientific investigation, be aware of the depth of thinking it will require. The table on the following page provides support in writing questions at different levels of thinking or complexity. It also provides some key words to help target these thinking levels and gives some examples of questions.

When developing your question for investigation, be conscious of the level at which you are pitching it. Questions for scientific investigation should generally be at the analysis level.

Level of question complexity	Type of thinking	Words that might be used		Examples of questions and commands
Simple	**Retrieval:** remembering, producing information on demand	• who • where • list • show • describe • select • complete • define	• what • when • label • demonstrate • name • state • recognise • identify	**1** Define the term 'diffusion'. **2** What is the function of an enzyme? **3** Identify the structural components of a nucleotide. **4** What are the products of photosynthesis?
	Comprehension: the ability to understand information	• who • where • explain • represent • show how	• what • when • summarise • draw • describe	**1** Represent the process of osmosis with a labelled diagram. **2** Why are the processes of photosynthesis and cellular respiration described as inverse reactions? **3** Explain how enzymes may be described using the 'lock-and-key' model. **4** What happens to enzymes when they are subjected to extremes of pH?
	Application: using knowledge in new situations, including: • testing a hypothesis • solving a problem • experimenting and using data • decision-making	• why • investigate • find out about • test • solve • develop • decide • construct	• how • research • experiment • predict • adapt • judge	**1** Many scientists believe that limiting human population growth is necessary to control environmental damage. Construct an argument for or against this statement. **2** Investigate the effect of vitamin B12 deficiency in humans. **3** Research the link between rising global temperatures and coral bleaching on the Great Barrier Reef.
Complex (requires more thinking)	**Analysis:** scrutinising and breaking something into its smaller parts, including: • comparing • classifying • identifying errors • concluding • predicting • judging	• why • categorise • contrast • sort • organise • generalise • evaluate • edit • assess • judge	• how • compare • distinguish • discriminate between • deduce • critique • diagnose • identify errors • identify misunderstandings	**1** Organise the following in order from simplest to most complex: organ, specialised cell, system, tissue. **2** Compare open and closed transport systems in animals. **3** Compare and contrast physical and chemical digestion. **4** Organise the structures of the mammalian cardiovascular system from those with high blood pressure to those with low blood pressure. Present this information in a diagram or graph.

ISBN 978 0 6557 0026 5

SECONDARY-SOURCED INVESTIGATION

Some scientific investigations require the collation and analysis of data that others have collected. Data or information that was collected by someone else is known as secondary data or a secondary source. An investigation that uses secondary data is known as a secondary-sourced investigation. An example of a scientific investigation methodology that involves collating and analysing secondary sources is a literature review. This section guides you in conducting a secondary-sourced investigation.

Activities such as investigating an issue are likely to rely on secondary sources of information. Such investigations require you to think carefully about the topic; find, collect and organise information; analyse and synthesise findings; and present your ideas.

The process for undertaking a secondary-sourced investigation is summarised in the following flow chart. A secondary-sourced investigation is not necessarily a straightforward linear process, as shown in the flow chart. You can move back and forth between steps as needed.

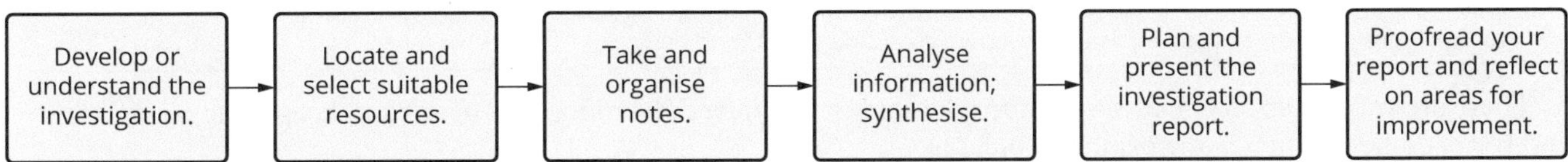

SOURCING INFORMATION

The resources you refer to in investigations may be primary and/or secondary.

Primary sources of information are created by a person directly involved in an investigation or study. Examples of primary sources are results from experiments (such as raw data or photographs), reports of scientific investigations, specimens or artefacts collected, and peer-reviewed scientific articles reporting the results of an investigation.

Secondary sources of information are a synthesis, review or interpretation of primary sources. Secondary sources include textbooks, biographies, documentaries, newspaper articles and websites.

Refer to the following two checklists. The first shows features to look for when assessing and selecting the best sources of information. The second shows how to set out the information that is required for the references section of your report in American Psychological Association (APA) seventh edition style. This is only one of several referencing systems that you might be required to use throughout your career.

Selecting resources for the investigation	Tick ✔
The resource is:	
• **credible** and I can identify the author, author's expertise and publisher	
• **current** because the date of publication of the material is provided and is recent	
• **factual** and I know that it is objective material and not biased	
• **accurate** and all information is correct	
• **relevant** and covers the area I am investigating	
• **readable** and neither too simple nor too complex in its coverage of the material.	

Examples of information required for references and bibliographies (APA style)
Article in scientific magazine Author, initials. (year). Title of article. *Journal title,* volume number(issue number), page numbers. Digital object identifier (DOI) or URL von Uexkull, N., & Buhaug, H. (2021). Security implications of climate change: A decade of scientific progress. *Journal of Peace Research, 58* (1), 3–17. https://doi.org/10.1177/0022343320984210
Book Author, initials. (year). *Title of book* (edition, if not first). Publisher. Bruns, E., Armstrong, Z., Bliss, C., Campanale, N., Georges, E., Hellberg, C., Jansen, C., Kerr, J., McMahon, K., Maginn, H., Meddings, J., Siwinski, S., & Ward, N. (2021). *Heinemann Biology 2* (6th ed.). Pearson Australia. Include names of all authors or editors up to 20. Special rules apply for 21 or more authors. You can find information about the rules online.
Internet Author, initials/name of organisation. (year). *Title of webpage or web document.* <URL> Royal Society of Biology. (2021). *Biology for beginners.* https://www.rsb.org.uk/

Always remember to record the above details for each resource you use. It is important to accurately cite resources in your reference section, as it can be time-consuming and difficult to find this information later.

NOTE-TAKING AND ORGANISING NOTES

Note-taking and organising requires skill. Good note-taking helps you avoid plagiarism and provides excellent information to support your scientific report writing. Plagiarism is when you take someone else's ideas and words and present them as your own work. You plagiarise if you copy sections or sentences from sources or you cut and paste from the internet. It is acceptable to use the ideas of others but you must state clearly where the information has come from in your reference section.

The following table provides examples of original text, plagiarised text and acceptably rephrased text.

Original text	Plagiarised text	Rephrased text
Dogs have sweat glands on their feet. Dogs pant when they are hot because their sweat glands are not sufficient to cool down their bodies. In addition, their tongues allow the water from their bodies to evaporate and cool down their bodies.	Dogs have their sweat glands on their feet. When dogs get hot they pant because their sweat glands are not enough to cool down their bodies. Their tongues let the water from their bodies evaporate and cool their bodies.	Dogs pant in order to cool down. Water evaporates from their tongues and this lowers their body temperature. They can't get cool enough through just the sweat glands on their feet.

There are various approaches to effective note-taking. Whatever technique you use, try to keep your notes brief and focus on key points. Some examples include:

- dot point summaries
- underlining or highlighting text
- labelled diagrams
- flow charts—to show sequences
- concept maps—to show connections between ideas
- Venn diagrams—to show similarities and differences
- tables—useful for summarising longer and more complex information that has subparts and can incorporate any of the other note-taking techniques. Adapt the table to suit your style and the task.

The following (partially completed) table shows how this technique can be used to take notes for a secondary-sourced investigation.

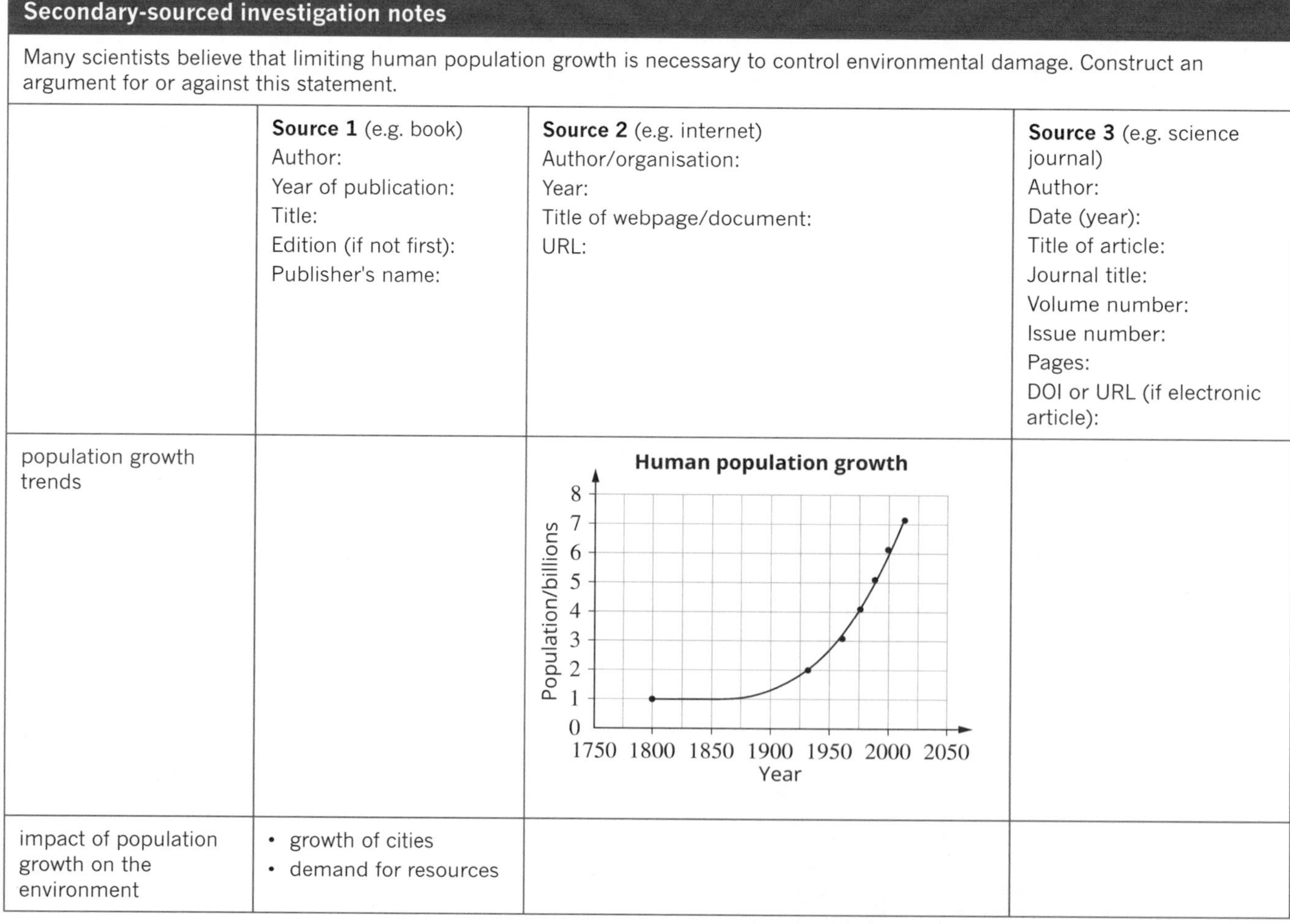

Secondary-sourced investigation notes			
Many scientists believe that limiting human population growth is necessary to control environmental damage. Construct an argument for or against this statement.			
	Source 1 (e.g. book) Author: Year of publication: Title: Edition (if not first): Publisher's name:	**Source 2** (e.g. internet) Author/organisation: Year: Title of webpage/document: URL:	**Source 3** (e.g. science journal) Author: Date (year): Title of article: Journal title: Volume number: Issue number: Pages: DOI or URL (if electronic article):
population growth trends		**Human population growth** (graph)	
impact of population growth on the environment	• growth of cities • demand for resources		

ISBN 978 0 6557 0026 5

Secondary-sourced investigation notes			
reasons to control and limit population growth	• increased demand for resources for human survival is unsustainable and causing environmental damage		
reasons not to limit population growth but to allow natural population growth			• other factors are contributing to environmental damage; land-use policies can be improved

SCIENTIFIC WRITING

Scientists have a particular writing style. You should use this distinctive style to communicate your ideas. Scientific writing is:

- objective—it describes events rather than what people think or feel and is as free as possible of bias or personal opinion
- precise—it avoids exaggeration and uses qualified language
- formal—it uses scholarly language rather than colloquial or everyday language
- concise—it conveys information in short, clearly understandable sentences without unnecessary information
- simple—it uses short sentences where possible
- predominantly written in passive voice, although sometimes active voice can be used to avoid confusion
- structured to include headings, tables, diagrams and mathematical calculations.

Examples of unscientific and scientific writing are demonstrated in the following table.

Unscientific writing	Scientific writing
Subjective, biased writing: • The results were fantastic. • This produced a disgusting odour. • The breathtakingly beautiful bowerbird…	**Objective, unbiased writing:** • The results showed… • This produced a pungent odour. • The golden bowerbird…
Exaggerated writing: • The object weighed a huge amount. • The magnesium burst into huge flames. • Millions of ants swarmed over…	**Accurate, precise writing:** • The mass of the object was 250 kg. • The magnesium burnt vigorously. • Ants swarmed all over…
Everyday, informal language: • The bacteria passed away. • The results don't… • We guessed that… • Previous researchers were slack and missed…	**Formal language:** • The bacteria died. • The results do not… • It was hypothesised that… • Previous researchers have not found…
Active voice: • We recorded oxygen levels every hour. • We put 50 g of solute in a conical flask containing distilled water, and then we slowly added 1 mol L^{-1} hydrochloric acid.	**Passive voice:** • The oxygen level was recorded hourly. • 50 g of solute was placed in a conical flask containing distilled water, and then 1 mol L^{-1} of hydrochloric acid was slowly added.

PRESENTING A REPORT ON A SCIENTIFIC INVESTIGATION

Scientific findings may be presented in a variety of ways. A common presentation format at science conferences is a poster. Posters can get ideas across to a large audience in an organised, concise and creative way. Other common presentation formats are essays, reports, oral presentations and articles. Each presentation format has its own conventions. The following table summarises the characteristics of a number of presentation formats.

Presentation format	Characteristics/inclusions	
poster	• balance of text and visuals • title, subheadings • balanced layout • captions for figures and tables	• references • hierarchy of font size according to subheading level • consistent font style—no more than three fonts
report/article	• structured with an introduction, paragraphs and conclusion • includes subheadings	• mainly text • can include diagrams, graphs and tables
essay	• structured with an introduction, paragraphs and conclusion • introduction states focus of essay • each paragraph makes a new point supported by evidence	• each paragraph links back to last paragraph • a text-style presentation format—visuals at end in appendix • conclusion draws all ideas together but does not include any new information
oral presentation	• needs to be engaging • refer to cue cards but do not read from them • watch audience as you speak	• stand still and avoid fidgeting • look at audience and appear confident

PROOFREADING

After you have completed the investigation and prepared your presentation, it is important to think about and check what you have done.

Proofread your work to minimise errors and maximise effective communication of the ideas from your investigation. Use the following questions as a proofreading checklist.

Proofreading checklist	Tick ✔
Have I:	
• investigated the question fully?	
• expressed myself clearly to communicate my ideas well?	
• used the scientific writing style?	
• included data analysis?	
• checked spelling, punctuation and grammar?	
• included references?	
• met the requirements of the presentation format?	

ISBN 978 0 6557 0026 5

Study skills

There are a variety of techniques and strategies you can use to help you study. You may find that you use different strategies in different situations. For example, you may prefer to highlight key phrases in your notebook throughout the year but make summaries of topics before an examination. The strategies you choose will depend on personal preference and may not be the same as those used by your classmates.

Effective study skills involve more than the learning strategies you use. Equally important is when you use those skills. It is more effective to apply study skills throughout the year, revising and consolidating your knowledge as you progress through the course, rather than doing a rushed cram just before the examination. Revise your work regularly. Being organised and setting up a study plan is key to reducing your stress.

GETTING ORGANISED

To get yourself organised, try the following steps.

- Use a diary to write down all homework and assessment tasks as soon as you get them. Note due dates and what is required.
- Be specific about the tasks you need to do. Rather than writing 'do biology', it is more effective to note things such as which questions to answer and which page to look at in your student book.
- Write a list of everything you need to do each day. Tick off or cross out items as you complete them.
- Break down larger tasks into smaller separate parts that are manageable.
- Make sure your lists and planners are realistic. Do not set yourself more than you can actually do.

STUDY TECHNIQUES

Studying requires concentration. Remove any distractions and factor in some breaks. Allow a 10-minute break every hour. Vary your study technique depending on the content to be learned and your personal preference. Although you may have already found a study technique that works for you, also consider the following options.

Study technique	Tips
Highlighting FUNGI Fungi often look like plants but do not use photosynthesis. Instead they feed on dead and decaying material, breaking it down further and helping chemical elements to return to the natural environment. Mushrooms, toadstools, yeasts and moulds are different types of fungi.	Highlight or underline key points as you read your notes or text.
Summary notes *ENZYMES* *Enzymes are:* • *composed of protein* • *substrate-specific* • *denatured by exposure to excessive heat* • *denatured by exposure to extremes of pH.*	• Create a list of key headings and add some dot points about each heading. • Write your own summary of the key ideas in each chapter. • Use headings and subheadings. • Underline key words and key phrases. • Use simple diagrams. • Remember, the most effective chapter summaries are clear, concise and uncluttered.
Diagrams chloroplasts thickened inner wall stomatal opening guard cells (turgid)	• Diagrams can be used as a summary of key concepts. • Diagrams are useful memory triggers. • Diagrams cover a lot of information in a visual way, with minimal text.

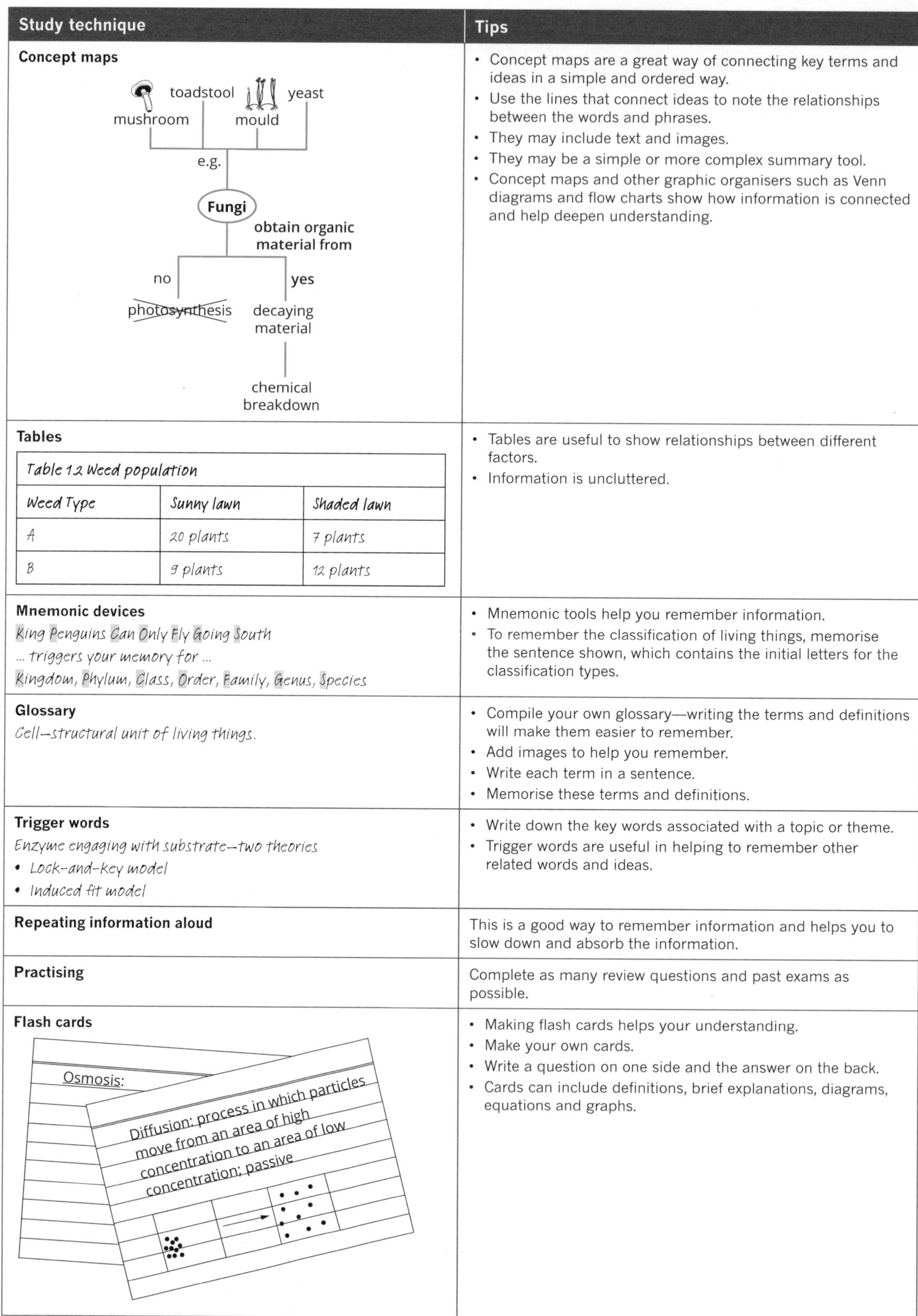

<table>
<tr><th>Study technique</th><th>Tips</th></tr>
<tr><td>Concept maps
</td><td>• Concept maps are a great way of connecting key terms and ideas in a simple and ordered way.
• Use the lines that connect ideas to note the relationships between the words and phrases.
• They may include text and images.
• They may be a simple or more complex summary tool.
• Concept maps and other graphic organisers such as Venn diagrams and flow charts show how information is connected and help deepen understanding.</td></tr>
<tr><td>Tables
<table>
<tr><th colspan="3">Table 1.2 Weed population</th></tr>
<tr><th>Weed Type</th><th>Sunny lawn</th><th>Shaded lawn</th></tr>
<tr><td>A</td><td>20 plants</td><td>7 plants</td></tr>
<tr><td>B</td><td>9 plants</td><td>12 plants</td></tr>
</table></td><td>• Tables are useful to show relationships between different factors.
• Information is uncluttered.</td></tr>
<tr><td>Mnemonic devices
King Penguins Can Only Fly Going South
... triggers your memory for ...
Kingdom, Phylum, Class, Order, Family, Genus, Species</td><td>• Mnemonic tools help you remember information.
• To remember the classification of living things, memorise the sentence shown, which contains the initial letters for the classification types.</td></tr>
<tr><td>Glossary
Cell—structural unit of living things.</td><td>• Compile your own glossary—writing the terms and definitions will make them easier to remember.
• Add images to help you remember.
• Write each term in a sentence.
• Memorise these terms and definitions.</td></tr>
<tr><td>Trigger words
Enzyme engaging with substrate—two theories
• Lock-and-key model
• Induced fit model</td><td>• Write down the key words associated with a topic or theme.
• Trigger words are useful in helping to remember other related words and ideas.</td></tr>
<tr><td>Repeating information aloud</td><td>This is a good way to remember information and helps you to slow down and absorb the information.</td></tr>
<tr><td>Practising</td><td>Complete as many review questions and past exams as possible.</td></tr>
<tr><td>Flash cards
</td><td>• Making flash cards helps your understanding.
• Make your own cards.
• Write a question on one side and the answer on the back.
• Cards can include definitions, brief explanations, diagrams, equations and graphs.</td></tr>
</table>

ISBN 978 0 6557 0026 5

Study technique	Tips
Teaching someone	• Teach friends or family members. • Teaching a difficult concept to someone means you must first understand the concept yourself.
Handwriting notes	• Handwrite rather than type summary notes. • Remember, the examination requires you to write answers. • Practise writing for long stretches of time and make sure your writing is legible.
Responding to feedback and self-correcting	• Check through all feedback from your teacher. • Highlight what was right or wrong. • Attempt to identify where you have errors and rework the answer to get it right.

Examination preparation

In the weeks before the examination, begin your exam preparation. The earlier you begin revising, the easier it will be. It is also helpful to begin practising exam-style questions as early as possible, not just in the weeks before the exam.

Like most skills, practice will improve your ability to do exams and to handle different types of exam questions. Doing practice exams is vital because you gain experience in:

- using reading time effectively
- allocating the right amount of time to each question
- working to a time limit
- reading and interpreting questions
- understanding what is required for each question
- planning answers
- deciding on relevant information
- proofreading/checking your answers
- writing efficiently for the duration of the exam.

Use your practice exam experience not only to revise but also to analyse your strengths and weaknesses, and to assess how you managed your time.

Use the following checklist as a reminder of your study program.

Study program checklist	Tick ✓
Have I:	
• revised all areas of the course?	
• highlighted important points?	
• made a summary of the important points in each topic?	
• read over my revision notes?	
• looked at and worked through practice exam papers?	
• answered practice exam questions within the appropriate time limit?	

EXAMINATION STRATEGIES

Familiarise yourself with the conditions of the examination well before the day you sit the exam. You should know:

- the number of exams for the subject
- the amount of reading time allowed in the exam before writing begins
- the amount of writing time allocated
- any particular equipment allowed and/or required, such as a calculator, pencils, pens and ruler
- strategies to tackle the exam.

Exam strategies are listed in the following table.

Exam strategies

Reading time

- You will be given reading time at the beginning of the exam (usually 15 minutes).
- Remember that no writing at all is allowed during this time—no note-taking, no highlighting or underlining.
- Read the instructions.
- Read through the short-answer questions first—this will give you an overall sense of the themes of questions that require written responses.
- Read the multiple-choice questions next.
- Decide the order in which you will answer questions. Start with what you consider to be the easiest question to build confidence.

Writing time

- You will be given 2.5 hours of working time.
- Begin with the multiple-choice questions.
- Answer every multiple-choice question, even if you can only make an educated guess.
- If you are unsure of an answer to a multiple-choice question, mark it so you can come back to it if time allows.
- Attempt the short-answer questions next.
- Attempt the easiest short-answer questions first and work your way to the more challenging questions.

Tips for answering questions

- Carefully read each question, underlining key words.
- Be aware that most questions are structured so they become more challenging towards the end. You may not be able to answer the last part of a question but you will have earned most of the marks by answering earlier parts.
- Look carefully at any diagrams, pictures, tables and graphs and make sure you understand their relevance to the questions involved.
- For questions with graphs, read the graph title and the labels on the axes carefully so that you can establish the relationship the graph is showing.
- For questions with tables, read the headings on the columns and rows carefully so that you can analyse the content of the table effectively.
- Check for key words in a question. Highlight them but do not colour the whole question.
- Use correct spelling—it can mean the difference between scoring a point and not scoring a point, for example, 'miosis' could be misspelling for either 'mitosis' or 'meiosis' and consequently would score zero.
- Plan your answers before you write, remembering to address the exam criteria.
- For questions with parts, read the whole question first. This gives you an overall picture of the question. It will also help to ensure that you do not repeat yourself in subsequent parts of the question.
- Make sure you actually answer the question that is asked.
- Once you have answered the question, re-read your answer and then re-read the question to ensure that you have actually answered all of the question.
- When writing a definition, avoid using the word you are defining in your definition.
- If giving values from a graph, use a ruler to line up points with the axes so you can be accurate, and always include units in your answer.
- Be sure to attempt all questions.
- Read over your answers to pick up careless errors—the mind is faster than the hand and you may not always write what you intend (especially when you have limited time).
- Write legibly. Exams are scanned and marked online. If the assessor can't read your answer, they can't mark it as correct.
- Keep an eye on the time.
- Never leave an exam early. Use any spare time to re-read and check your answers.

Exam cues

- The number of marks allocated to a question provides a clue about how much you are expected to write. Two marks usually means you need to make a minimum of two points.
- The number of lines allowed for the answer indicates the length of the expected answer. If your writing is large, you may need to turn the page over and continue on the back. Make sure you indicate to the examiner that they must turn to the back of the page for the rest of the answer. Where the answer continues, clearly state that it is the continuation of the question and state the question number.

ISBN 978 0 6557 0026 5

HANDLING EXAM QUESTIONS

The exam is divided into two sections. Section A consists of multiple-choice questions and Section B contains a series of short-answer questions.

Multiple-choice questions	Short-answer questions
• In many cases you can use your basic knowledge of the subject to cancel out one or two options. • Mark an answer to every question on your answer sheet as you go through them. This avoids leaving any questions unanswered if you run out of time, and avoids getting out of order by leaving blank rows. You can always go back and change the answers you were unsure about. • Circle the letter of your answer to the question on the exam paper as well, so you can easily check if you get out of sync on the answer sheet. • If you are finding a question difficult, do not spend too much time on it as it is only worth 1 mark. Make your best educated guess, if necessary, and then mark the question on your paper and come back to it later if you have time. • If you have time at the end of the exam, redo the questions with your original answer covered and compare.	• Be careful to do what the question asks. For example, do you need to identify, describe, explain or compare? Each requires a different response. • When in doubt, point it out, that is, give a thorough answer. Dot points and labelled diagrams are acceptable as long as you fully address the question. • Tailor your answer specifically to the question being asked. Do not rely on a general answer when given a specific scenario. For example, indicate by what mechanism the population decreased, do not just say it decreased. • Look at the marks allocated for an indication of the level of detail required. Generally 1 mark = 1 key point, 2 marks = 2 key points, etc. • Give full explanations and show your working. This makes it easier for you to check your answer and you often gain part marks and consequential marks if the marker can see where you made an error.

Every year, examiners comment that many students fail to answer the question: that is, students' answers are 'not relevant' or are 'off the point'. Giving a relevant answer to a question is vital and depends on:

- choosing appropriate content
- using content purposefully to do what is asked (see the key action words list below).

Key action words	
analyse	break down into key points; show the essence of; inquire into; identify components and the relationship between them; draw out and relate implications
compare	show points of similarity
contrast	show points of difference
describe	give the features/characteristics of
design	provide steps for an experiment or procedure
evaluate	make a judgement based on criteria; determine the value of
explain	provide details to give the reader an understanding of; relate cause and effect; make the relationships between things evident; provide why and/or how
justify	demonstrate the correctness of (by giving evidence); give evidence in support of an argument or conclusion
suggest	put forward ideas, proposals or recommendations

EXAMPLES OF EXAM RESPONSES

Each year the examiners write an assessment report on the most recent exam. Reading these reports is a good way to understand what the examiners expect in the answers. The reports indicate where students showed strong responses and where answers could be improved.

Looking closely at sample responses will help you identify what is required to gain full marks. Two sample responses are provided below: one is well done (high-level response), while the other has room for improvement (low-level response). The annotations draw your attention to key points to note in each response.

High-level response

Question 1 (12 marks)

a DNA replication occurs during the synthesis phase of the cell cycle. This is followed by nuclear and cell division. Describe the type of cells that result from mitosis and meiosis, shown below, and complete the labels in the diagrams provided.

i mitosis

ii meiosis

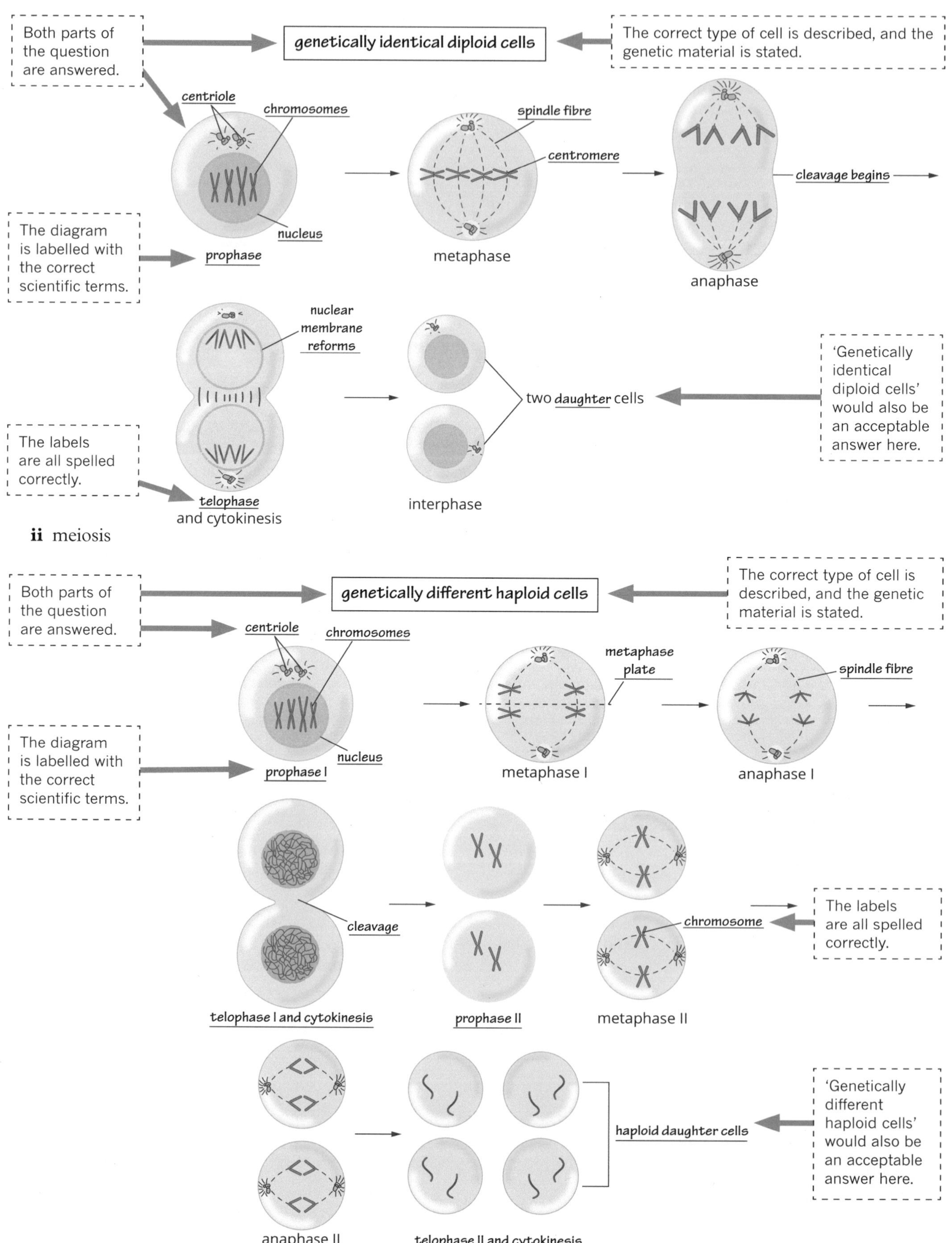

ISBN 978 0 6557 0026 5

b Consider the image of two homologous chromosomes shown to the right.

Name the point at which these chromosomes cross over, and identify the stage named in part **a** in which this event occurs. Explain the significance of this event for the continuity of species.

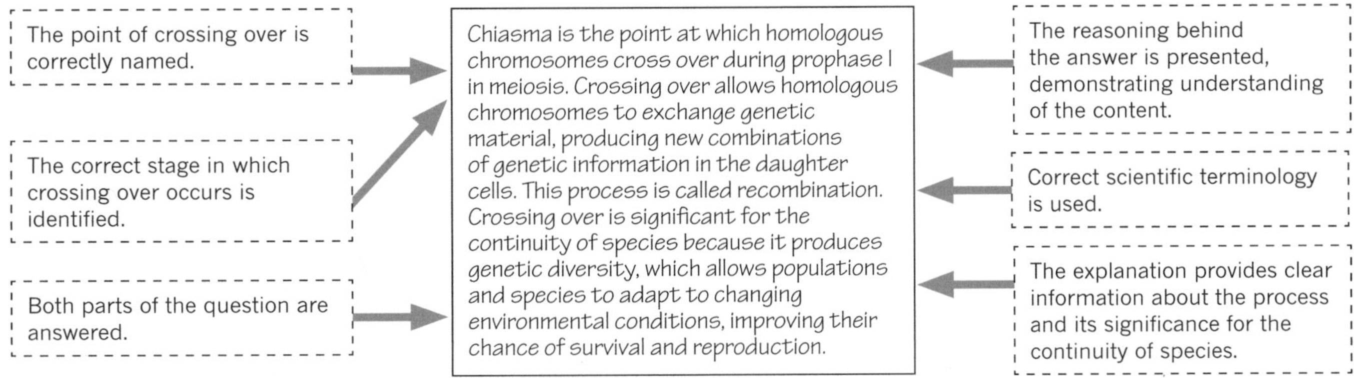

c Identify and explain:

i one disadvantage of asexual reproduction

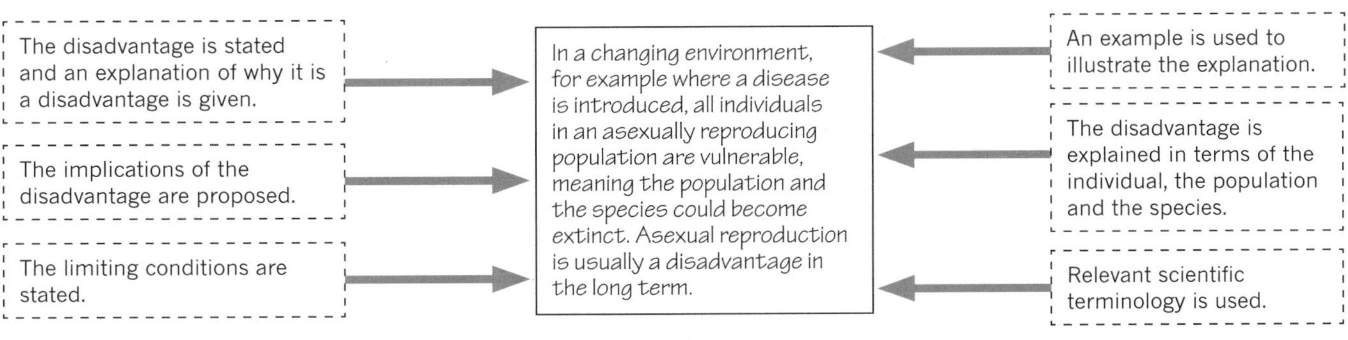

ii one advantage of sexual reproduction.

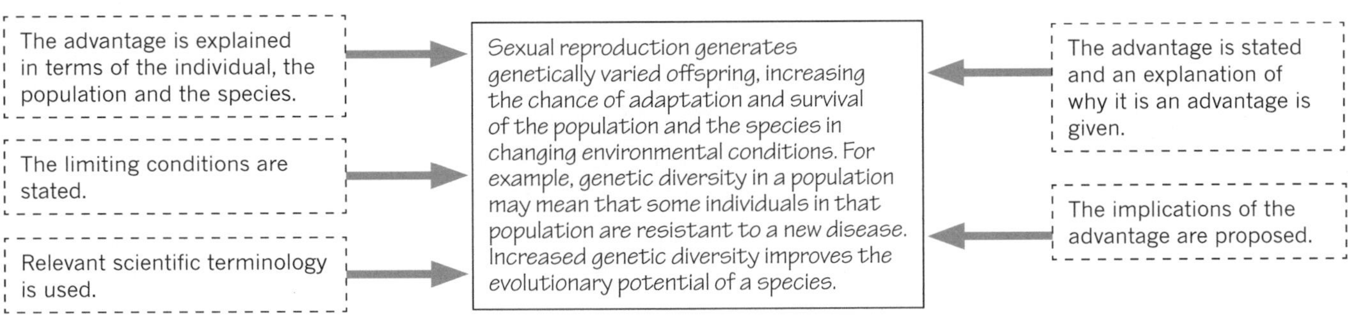

Low-level response

Question 1 (12 marks)

a DNA replication occurs during the synthesis phase of the cell cycle. This is followed by nuclear and cell division. Describe the type of cells that result from mitosis and meiosis, shown below, and complete the labels in the diagrams provided.

i mitosis

ii meiosis

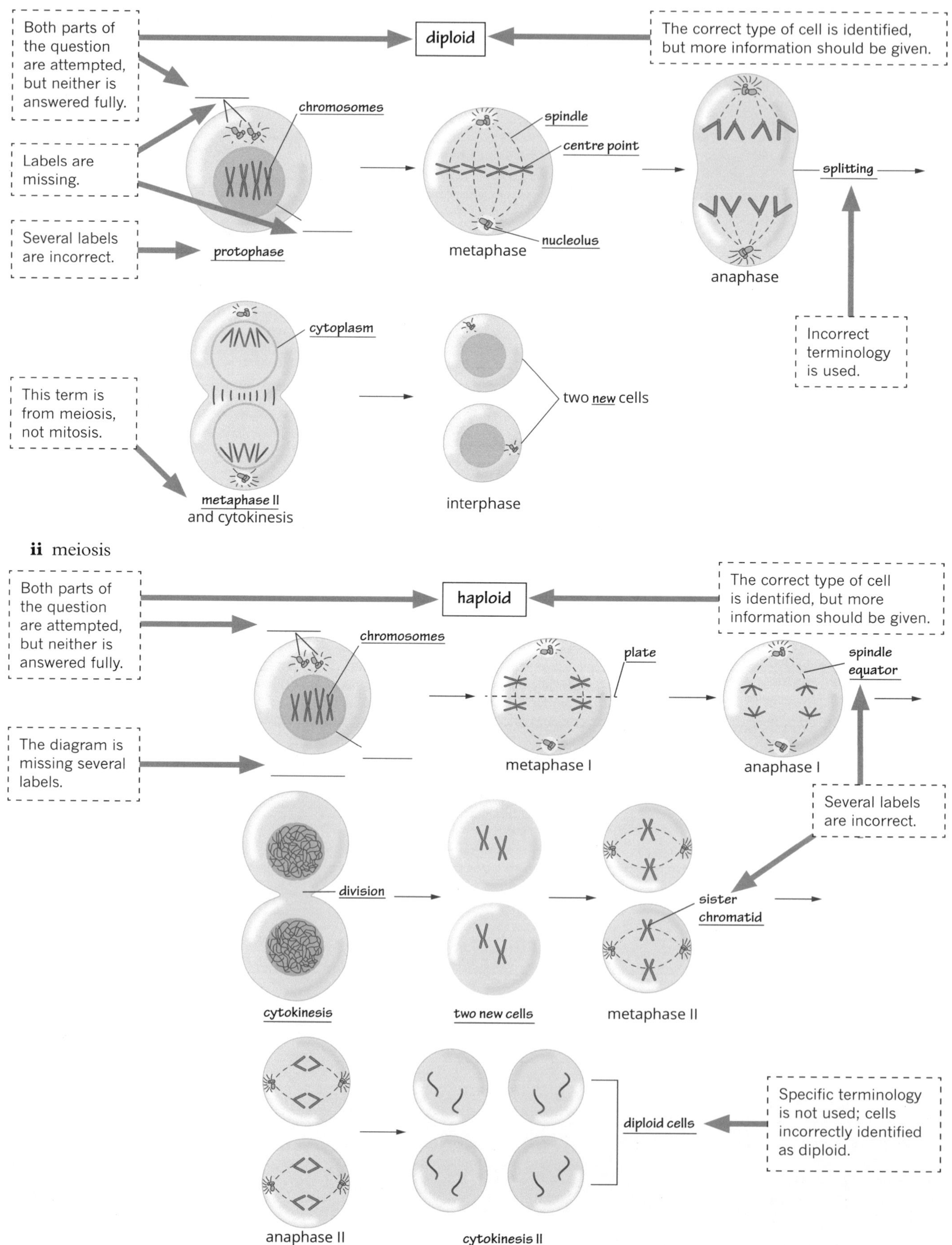

 ISBN 978 0 6557 0026 5

b Consider the image of two homologous chromosomes shown to the right.

Name the point at which these chromosomes cross over, and identify the stage named in part **a** in which this event occurs. Explain the significance of this event for the continuity of species.

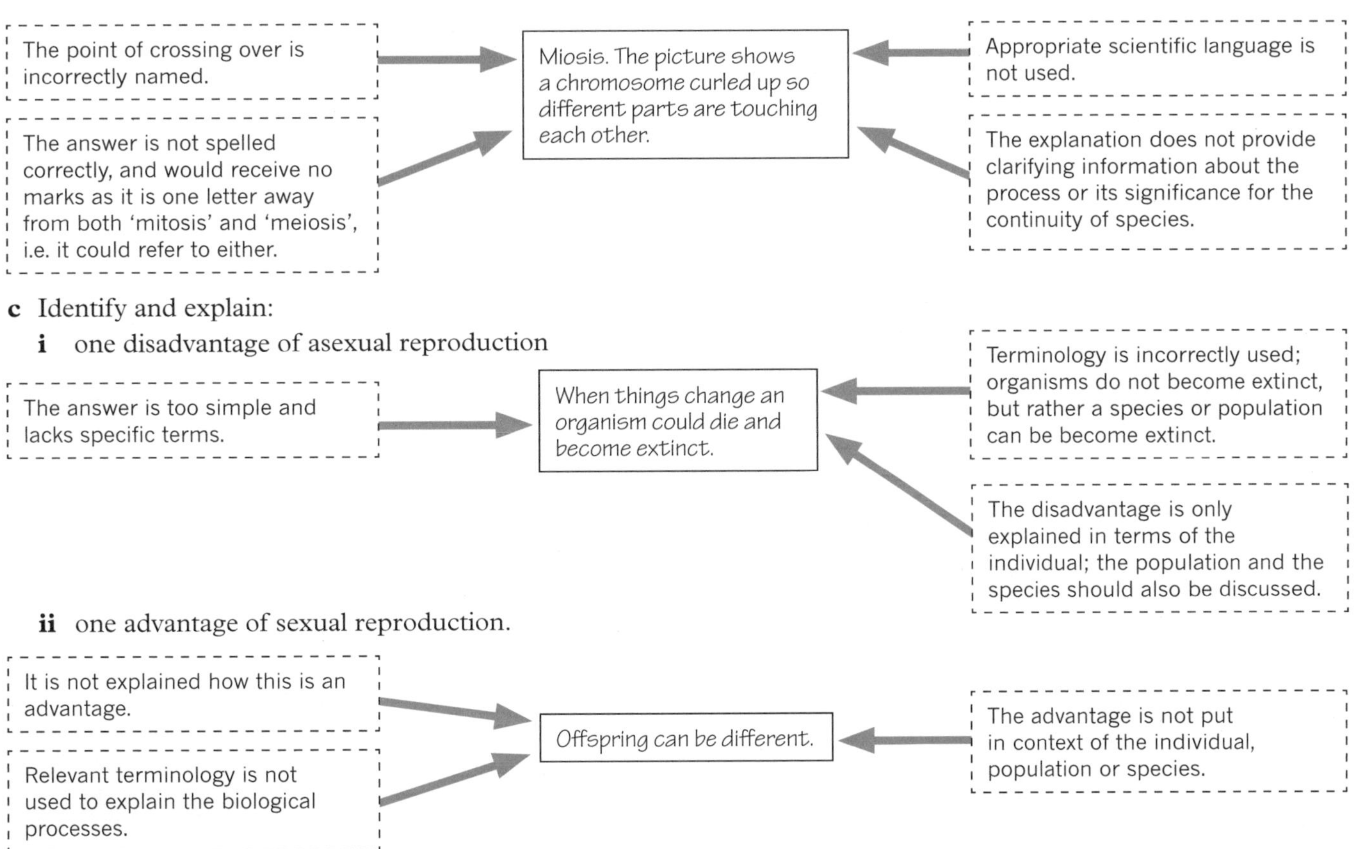

c Identify and explain:

i one disadvantage of asexual reproduction

ii one advantage of sexual reproduction.

It is not explained how this is an advantage.

Relevant terminology is not used to explain the biological processes.

Offspring can be different.

The advantage is not put in context of the individual, population or species.

GETTING READY

Once you have finished revising and feel confident with the material, use the following checklist to make sure you are prepared before the exam day arrives.

The exam itself	Tick ✔
Have you:	
• checked the exact exam requirements?	
• obtained and carefully read the instruction cover sheet from previous exams? (Check that no changes are expected for your exam.)	
• planned how to best use your reading time?	

Practical steps	Tick ✔
Have you:	
• assembled all materials required and allowed? This includes:	
- pens to write your answers	
- pencils of the correct lead for the multiple-choice answer sheet	
- sharpeners, erasers, rulers and other permissible materials, such as calculators and spare batteries	
- a water bottle (Check that your bottle meets the requirements—it may need to be clear with no label.)	
- a watch to keep track of the time (Mobile phones are not permitted.)	
• double-checked dates, times and locations for each exam?	
• remembered your student number?	
• packed a form of photo ID (if required)?	
• arranged transport, allowing ample time to avoid rushing and to manage possible delays? (Be sure to arrive early to settle your nerves and gather your thoughts.)	
• planned for appropriate meals, snacks and drinks? (Exam times vary. Eating heavy meals before exams can make you sleepy, so avoid doing this.)	

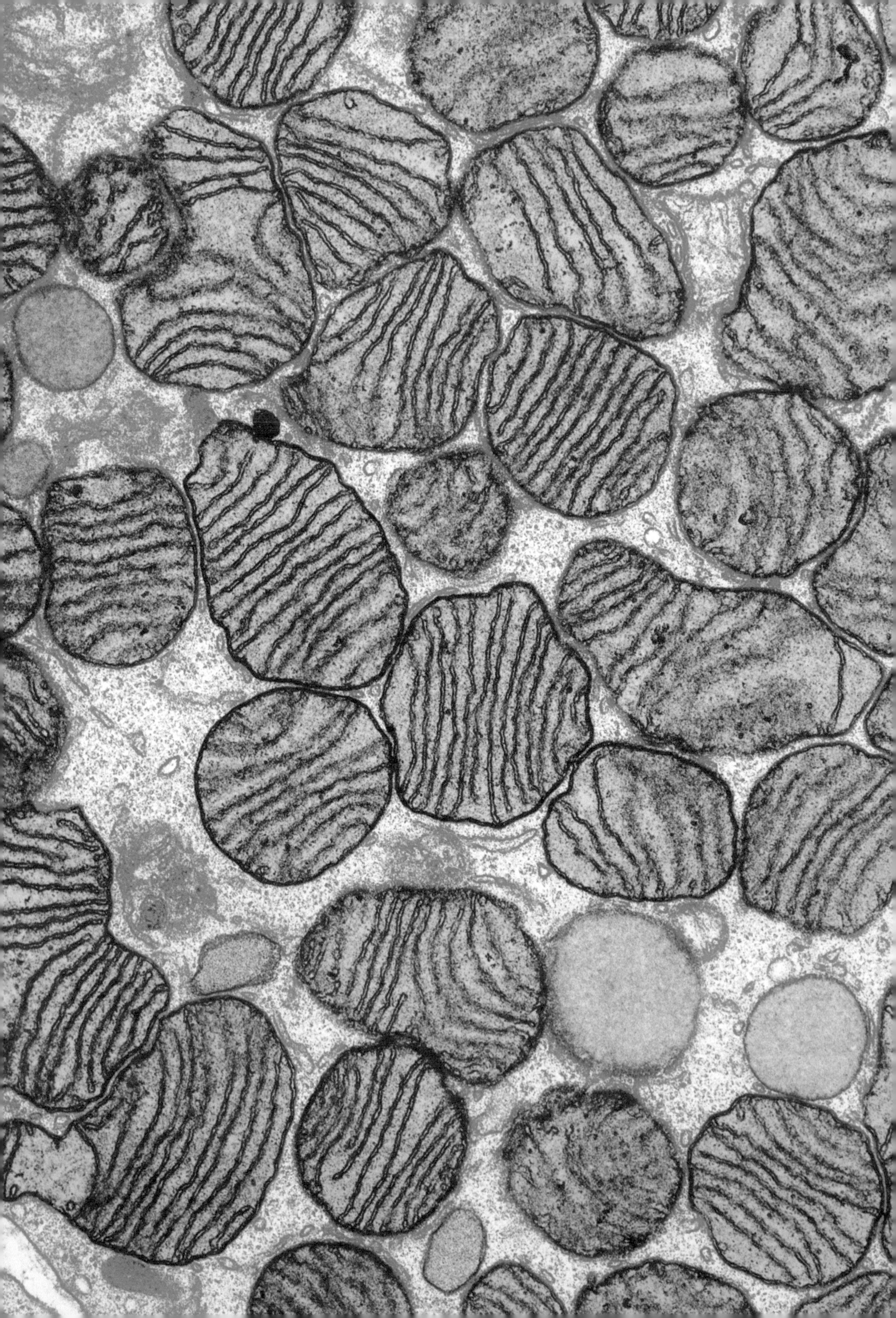

UNIT 3

How do cells maintain life?

AREA OF STUDY 1

What is the role of nucleic acids and proteins in maintaining life?

Outcome 1

On completion of this unit the student should be able to analyse the relationship between nucleic acids and proteins, and evaluate how tools and techniques can be used and applied in the manipulation of DNA.

Key knowledge

The relationship between nucleic acids and proteins

- nucleic acids as information molecules that encode instructions for the synthesis of proteins: the structure of DNA, the three main forms of RNA (mRNA, rRNA and tRNA) and a comparison of their respective nucleotides
- the genetic code as a universal triplet code that is degenerate and the steps in gene expression, including transcription, RNA processing in eukaryotic cells and translation by ribosomes
- the structure of genes: exons, introns and promoter and operator regions
- the basic elements of gene regulation: prokaryotic *trp* operon as a simplified example of a regulatory process
- amino acids as the monomers of a polypeptide chain and the resultant hierarchical levels of structure that give rise to a functional protein
- proteins as a diverse group of molecules that collectively make an organism's proteome, including enzymes as catalysts in biochemical pathways
- the role of rough endoplasmic reticulum, Golgi apparatus and associated vesicles in the export of proteins from a cell via the protein secretory pathway

DNA manipulation techniques and applications

- the use of enzymes to manipulate DNA, including polymerase to synthesise DNA, ligase to join DNA and endonucleases to cut DNA
- the function of CRISPR-Cas9 in bacteria and the application of this function in editing an organism's genome
- amplification of DNA using polymerase chain reaction and the use of gel electrophoresis in sorting DNA fragments, including the interpretation of gel runs for DNA profiling
- the use of recombinant plasmids as vectors to transform bacterial cells as demonstrated by the production of human insulin
- the use of genetically modified and transgenic organisms in agriculture to increase crop productivity and to provide resistance to disease.

REVISION

Cells

CELL THEORY

The cell theory is a foundation stone in the study of Biology. The cell theory states that:

- all organisms are made up of cells
- new cells are produced from existing cells
- the cell is the smallest organisational unit of a living thing.

There are two different kinds of cells—prokaryotic (found in bacteria and archaea) and eukaryotic (found in protists, fungi, plants and animals) (Table 3.1.1).

Differences also occur within the group of eukaryotic organisms. A typical plant cell has cell walls made from cellulose that provide structural support, chloroplasts that are the site of photosynthesis, and large vacuoles. These features are not present in animal cells (Figure 3.1.1 a and b).

Table 3.1.1 Cells and their features

Type of cell	Feature	Example
eukaryotic	distinct organelles such as ribosomes as well as membrane-bound organelles including nucleus, mitochondria, endoplasmic reticulum, Golgi apparatus, lysosomes, vesicles	organisms in Kingdoms Protista, Fungi, Plantae, Animalia, but not Bacteria and Archaea
prokaryotic	lack membrane-bound organelles; nuclear material present as a single, circular thread of DNA; plasma membrane surrounded by cell wall of protein and complex carbohydrate; relatively small cells; feature circular threads of DNA called plasmids	organisms in Kingdoms Bacteria (bacteria, cyanobacteria) and Archaea (includes some extremophiles)

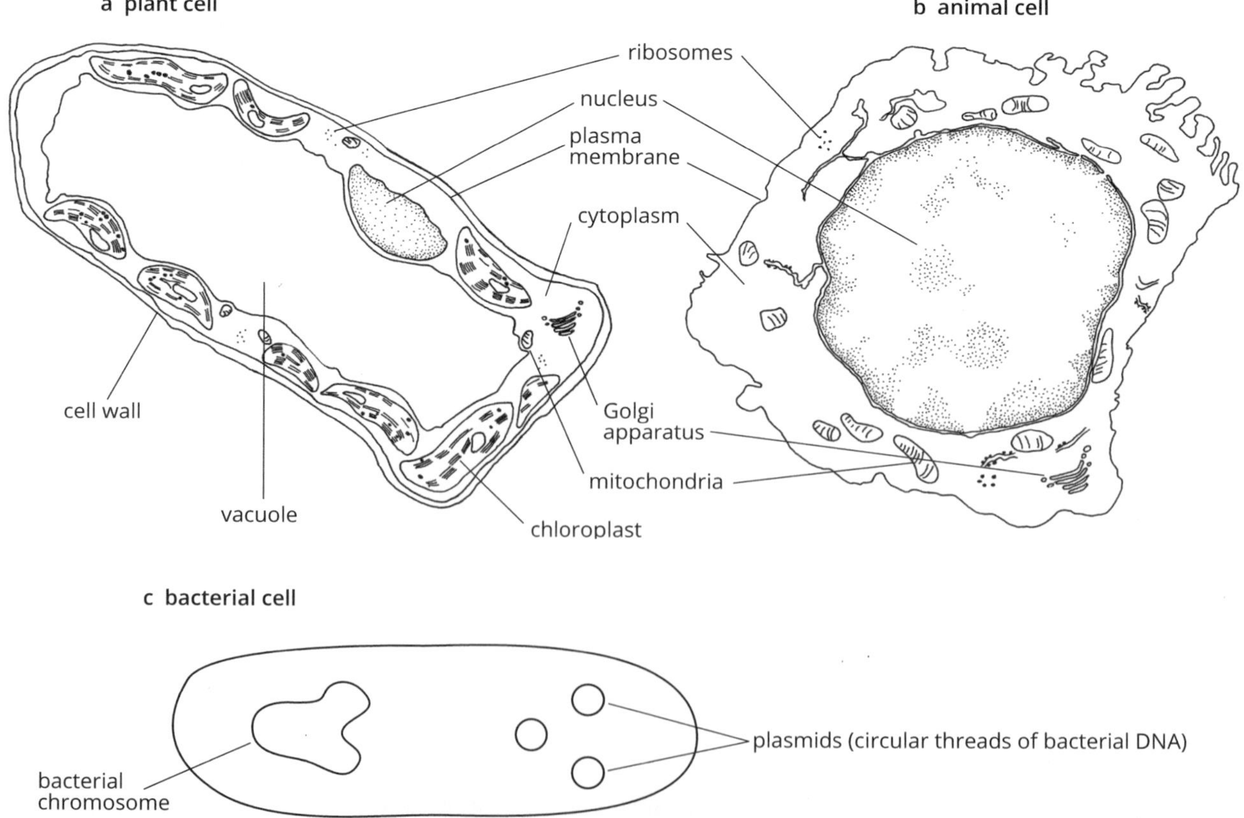

Figure 3.1.1 (a) Plant cell, (b) animal cell and (c) bacterial cell

REVISION

ORGANELLES

Organelles are subcellular structures with specialised functions (Table 3.1.2).

Table 3.1.2 Organelles and their functions

Organelle	Function
nucleus	cell reproduction and control of cellular activities; contains DNA
ribosomes	protein production; proteins produced by free ribosomes are used within the cell
mitochondria	aerobic cellular respiration
endoplasmic reticulum (ER)	transport of materials such as proteins and carbohydrates within cell; involved in synthesis of substances, e.g. protein, lipids, glycogen
rough endoplasmic reticulum (RER)	endoplasmic reticulum with associated ribosomes; ribosomes synthesise proteins and are transported throughout the cytoplasm on ER; proteins produced on ribosomes associated with endoplasmic reticulum are exported from the cell
Golgi apparatus	protein production completed; proteins packaged for dispatch from cell
cell wall	structural support in plants
plasma membrane	encloses cell contents; regulates passage of materials into and out of cell
cytoplasm	reservoir containing cell contents including water, ions, dissolved nutrients, enzymes and organelles
lysosomes	contain enzymes responsible for breakdown of debris
vacuole	storage facility for fluid, enzymes and nutrients
chloroplast	photosynthesis

Molecular composition of organisms

The cells that make up living organisms are themselves composed of key chemical elements. Those that occur in greatest proportion include carbon, hydrogen, oxygen and nitrogen. These elements combine to form a variety of important biomacromolecules. Biologically important molecules can be grouped into two types—organic and inorganic (Figure 3.1.2).

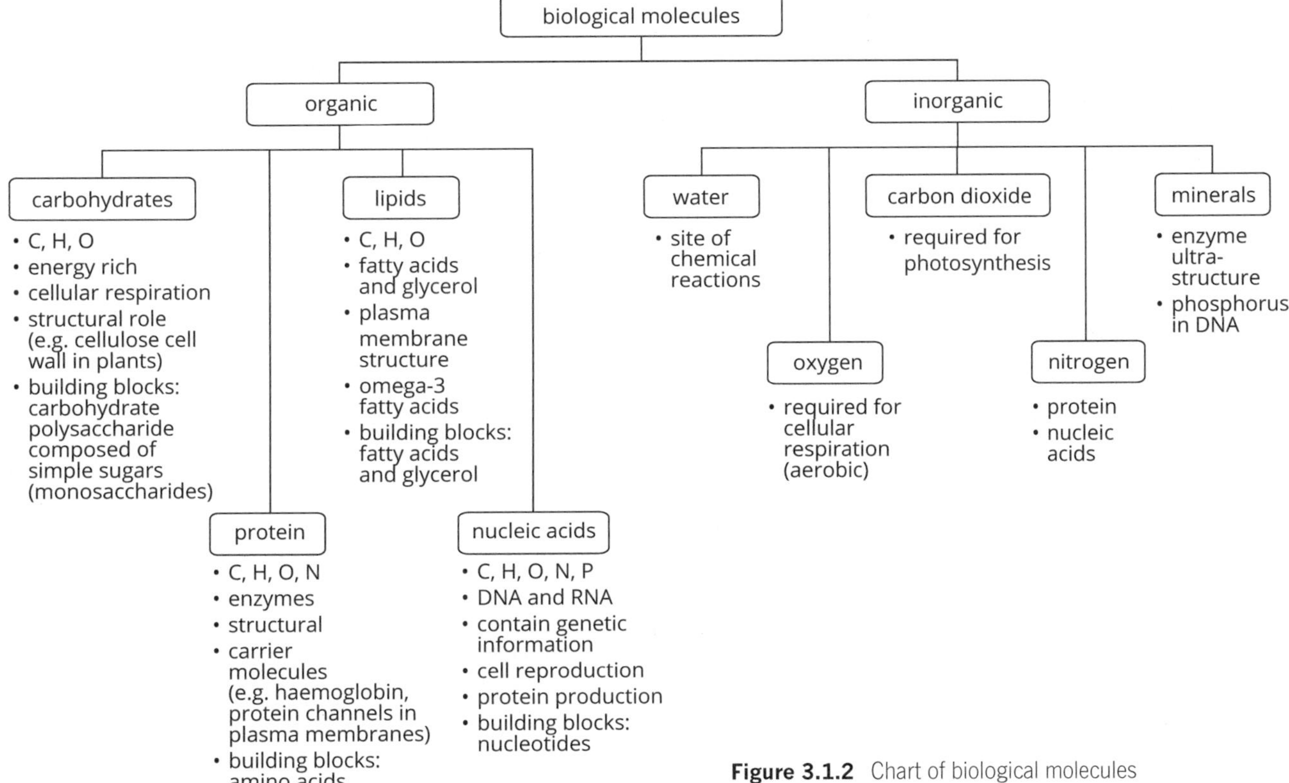

Figure 3.1.2 Chart of biological molecules

KEY KNOWLEDGE

REVISION

BIOMACROMOLECULES

Biomacromolecules are formed in condensation reactions. This involves a chemical reaction in which two smaller organic compounds are combined to form a larger organic compound and water is a by-product of the reaction (Figure 3.1.3). When polymers are constructed in this way the reaction is called condensation polymerisation.

This is true for the synthesis of all biomacromolecules, including lipids and polymers of carbohydrates, proteins and nucleic acids.

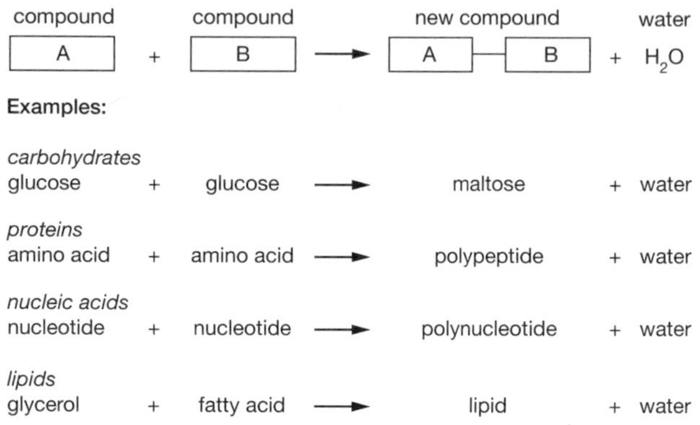

Figure 3.1.3 Some possible condensation polymerisation reactions involved in the formation of biomacromolecules

The relationship between nucleic acids and proteins

Common to all living things on Earth is the presence of the genetic material **deoxyribonucleic acid (DNA)** (Figure 3.1.4a). The DNA molecule stores the inherited characteristics of organisms and is transmitted to subsequent generations through reproductive processes in every species.

NUCLEIC ACIDS—DNA AND RNA

The structural unit of DNA is the **nucleotide** (Figure 3.1.4b). Nucleotides are named according to the base that each includes. The bases in DNA occur in complementary pairs:

Adenine (A) pairs with **thymine (T)**

Guanine (G) pairs with **cytosine (C)**

Adenine (A) and guanine (G) have a double-ring structure and are known as **purines**, while cytosine (C) and thymine (T) have a single-ring structure and are known as **pyrimidines** (Table 3.1.3). Nucleotides are chemically bonded together to form polymers called **nucleic acids**. Nucleic acids are essentially information molecules that contain the coded instructions for **protein synthesis**. The sequence of nucleotides in DNA is significant because of its role in protein production. A specific DNA sequence that codes for a particular protein is called a **gene**. The **genome** is the total complement of all of the genes in an individual organism. The **proteome** is the full complement of proteins in an individual. An organism's genome is intrinsically linked to its proteome.

The complementary strands of the DNA molecule are described as **antiparallel** because one runs 5' → 3' while the other runs 3' → 5'. Figures 3.1.4c and 3.1.5 show a simplified representation of a molecule illustrating this antiparallel arrangement.

Figure 3.1.4a The DNA in the nucleus of cells unravels to reveal a double helix.

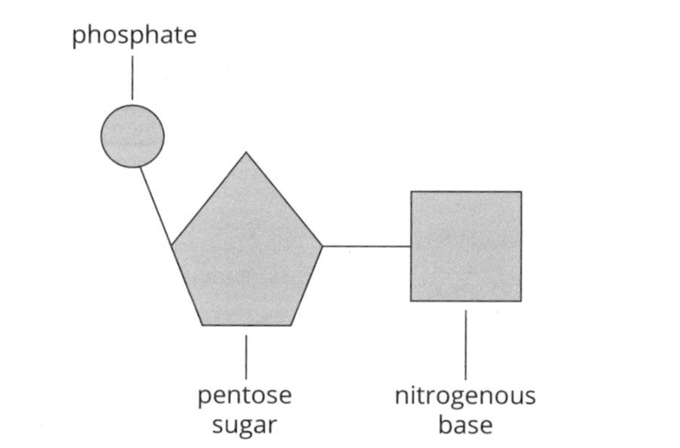

Figure 3.1.4b Nucleotide

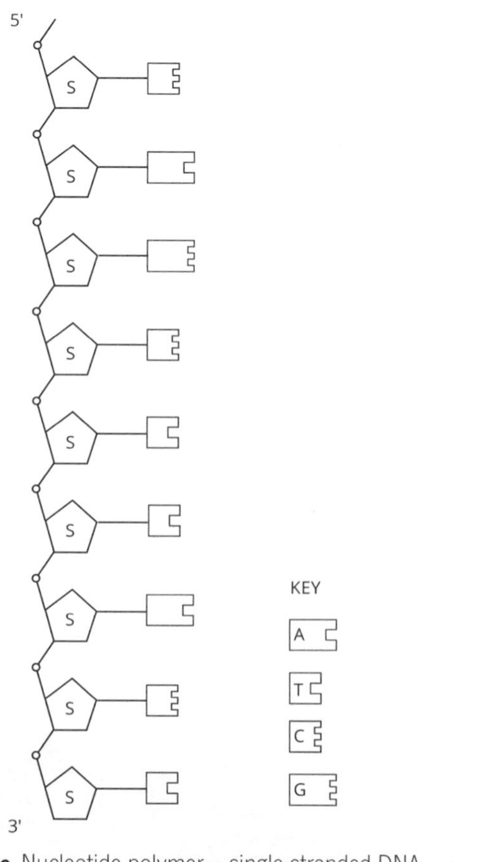

Figure 3.1.4c Nucleotide polymer—single-stranded DNA

 ISBN 978 0 6557 0026 5

KEY KNOWLEDGE

Table 3.1.3 DNA features

	Feature	Examples
Purines	double-ring structure	adenine, guanine
Pyrimidines	single-ring structure	thymine, cytosine

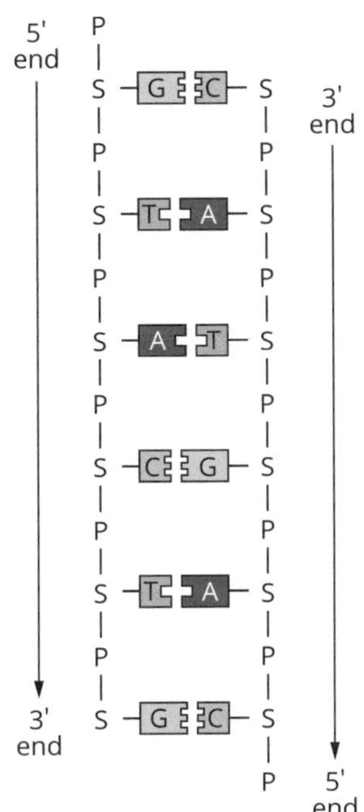

Figure 3.1.5 Nucleotides arranged in complementary pairs

Nucleic acids occur in two forms—DNA, which contains the genes, and **ribonucleic acid (RNA)** (Table 3.1.4). There are three kinds of RNA. All nucleic acids are composed of nucleotides, however, while DNA contains the base thymine, in RNA thymine is replaced with **uracil**. DNA is a double-stranded molecule (Figure 3.1.6), while RNA is single-stranded.

Table 3.1.4 Role of nucleic acids

	Role
DNA	DNA contains the instructions for protein synthesis.
mRNA	Messenger RNA is a copy of the DNA template strand, which takes instructions to the ribosomes in the cytoplasm.
tRNA	Transfer RNA is the molecule that brings amino acids to ribosomes during protein synthesis.
rRNA	Ribosomal RNA is synthesised in the nucleolus and forms part of the structure of the ribosomes.

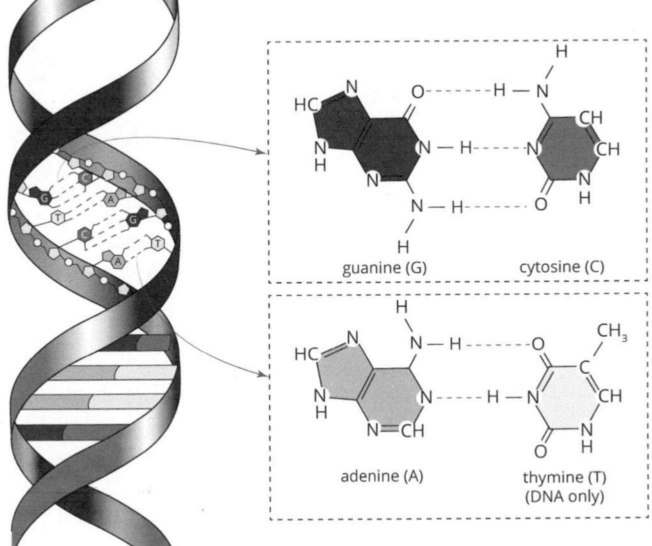

Figure 3.1.6 The helical structure of DNA. Two complementary strands form a double helix joined by base pairs guanine (G) and cytosine (C), and adenine (A) and thymine (T).

GENE EXPRESSION

The genetic information stored in DNA is used to synthesise the **amino acid** sequences that form **polypeptide chains** and then functional proteins through a process called **gene expression**.

The genetic code

The **genetic code** is a set of rules that defines how the information in nucleic acids (DNA and RNA) is translated into proteins and functional RNA molecules. The information in DNA and RNA is stored as a three-letter code of nucleotides. In DNA, this three-letter code is called a **triplet**. When a DNA triplet is transcribed into mature mRNA, the triplet is then called a **codon**.

Table 3.1.5 on page 6 shows the 20 amino acids of which polypeptides are composed, and the codons that code for each amino acid. A codon is composed of three nucleotide bases, hence the term 'triplet' code. To use this table, select the first base of the codon from the first column, read across the row for the second base, and then select the third base from the last column. The genetic code is considered 'universal' as it applies to almost all organisms on Earth. Because some codons code for the same amino acid, the genetic code is described as 'degenerate'.

KEY KNOWLEDGE

Table 3.1.5 Genetic code

First position (5' end)	Second position U	Second position C	Second position A	Second position G	Third position (3' end)
U	phe phe leu leu	ser ser ser ser	tyr tyr STOP STOP	cys cys STOP trp	U C A G
C	leu leu leu leu	pro pro pro pro	his his gln gln	arg arg arg arg	U C A G
A	ile ile ile met/ START	thr thr thr thr	asn asn lys lys	ser ser arg arg	U C A G
G	val val val val	ala ala ala ala	asp asp glu glu	gly gly gly gly	U C A G

Amino acid abbreviations	
ala	alanine
arg	arginine
asn	asparagine
asp	aspartic acid
cys	cysteine
gln	glutamine
glu	glutamic acid
gly	glycine
his	histidine
ile	isoleucine
leu	leucine
lys	lysine
met	methionine
phe	phenylalanine
pro	proline
ser	serine
thr	threonine
trp	tryptophan
tyr	tyrosine
val	valine

The sequence of amino acids represents the coded instructions for constructing polypeptides, the building blocks of proteins. The completed protein is the form in which the gene is expressed. Polypeptide production involves two key steps—**transcription** (Figure 3.1.7) and **translation** (Figure 3.1.8).

Transcription

- A DNA sequence is copied (transcribed) into a messenger RNA (mRNA) sequence.
- Transcription occurs in the nucleus.
- A DNA template strand is copied.
- mRNA is the single-stranded copy and contains the nucleotide base uracil instead of thymine.
- Transcription begins at the **promoter region** of a gene, a section of DNA that identifies the beginning of the gene.
- **Exons** are the coding regions of genes.
- **Introns** are the non-coding regions of genes.
- Both exons and introns are transcribed, forming a copy of the gene called pre-mRNA.
- Introns are subsequently cut out of the pre-mRNA, forming the final mRNA product.

Transcription process

DNA molecule unzips

↓

enzyme RNA polymerase moves along DNA template strand adding bases according to complementary base-pairing rules

↓

mRNA strand produced

At the conclusion of transcription, the mRNA molecule leaves the nucleus through nuclear pores and moves to the **ribosomes** where translation occurs.

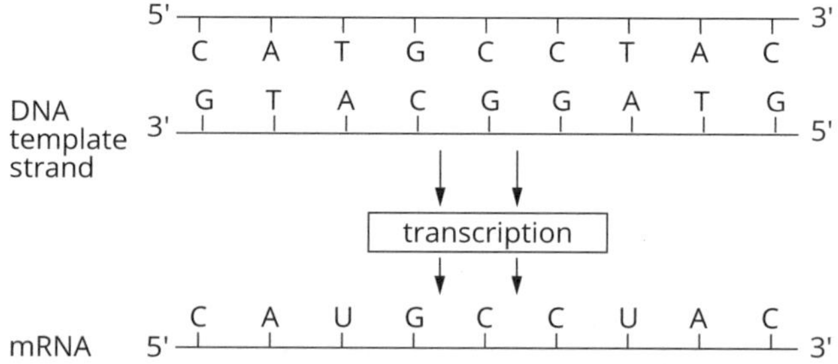

Figure 3.1.7 Transcription

 ISBN 978 0 6557 0026 5

KEY KNOWLEDGE

Translation

- A sequence of bases in mRNA is translated into an amino acid sequence (polypeptide).
- Each group of three bases (codon) in the mRNA codes for a particular amino acid—genetic code.
- Translation occurs in the cytoplasm at the ribosomes.

Translation process

ribosome moves along mRNA strand

↓

mRNA bases are read in groups of three called codons (in 5' → 3' direction)

↓

each codon specifies a particular amino acid (from the genetic code)

↓

tRNA molecule with complementary base triplet called an **anticodon** brings specified amino acid to ribosome

↓

amino acid monomers are linked with peptide bonds to form a chain called a polypeptide

Translation ceases when a **stop codon** is reached. The polypeptide chain is complete.

The polypeptide chain folds to form a complex three-dimensional structure and is then referred to as a protein.

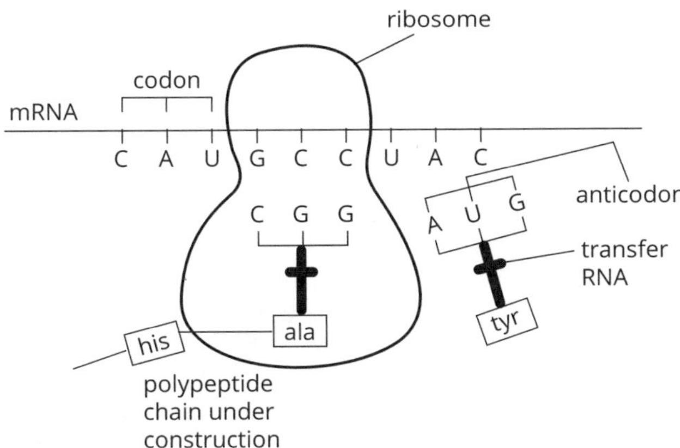

Figure 3.1.8 Translation

The processes of transcription and translation are summarised in Figure 3.1.9.

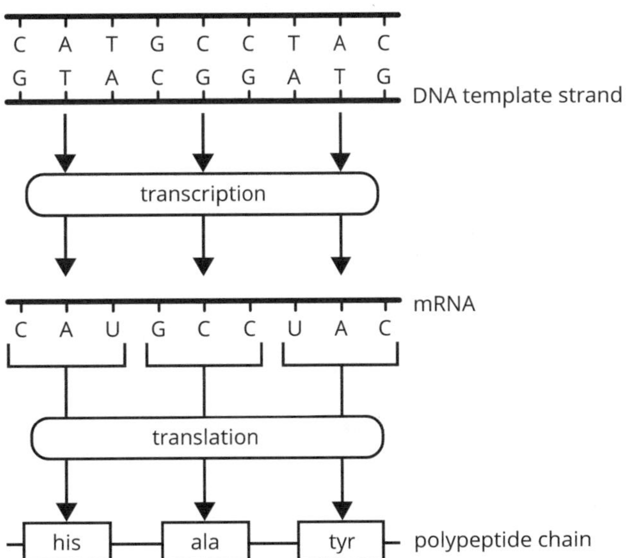

Figure 3.1.9 Summary of transcription and translation

GENE STRUCTURE AND REGULATION

Cells in the body show specialisation in structure and function, yet they all contain the same genetic information. This occurs as a result of the 'switching on and off' of different genes in particular cells. For example, beta cells in the pancreas have genes for insulin production switched on, but genes related to haemoglobin production are switched off. Gene regulation contributes to the conservation of energy and resources in cells.

Genes are typically structured so that the transcription of the coding region is carefully regulated. A **regulatory gene** (which controls transcription) precedes the **structural gene** (which codes for the protein). The regulatory gene is composed of a promoter region and an **operator region**, both upstream of the structural gene. The promoter region regulates when transcription should begin; the operator region effectively switches the gene on or off, allowing transcription to begin or cease. Proteins that bind with DNA at the promoter region to help regulate transcription are called **transcription factors**. Protein molecules that bind to DNA, thereby repressing or halting transcription, are called **repressor proteins**. The structural gene contains introns and exons. Once the gene is switched to on, transcription proceeds. Transcription is halted by the stop codon downstream of the structural gene (Figure 3.1.10).

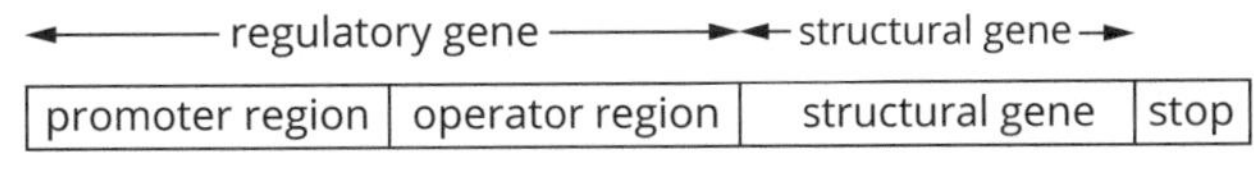

Figure 3.1.10 Regulatory gene

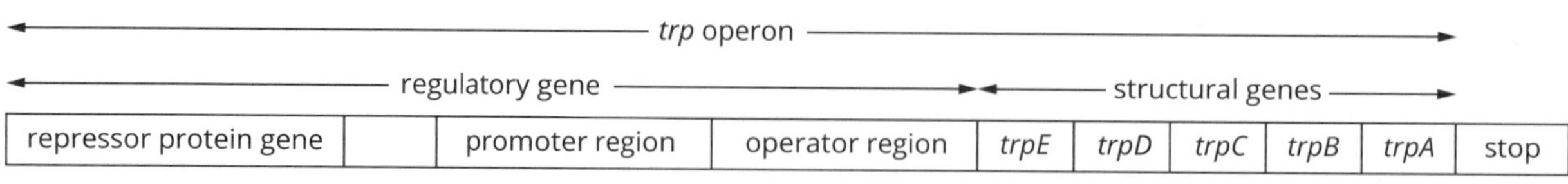

Figure 3.1.11 *trp* operon

The ***trp* operon** (tryptophan operon) model in bacteria serves as a classic example of our understanding of gene function. An **operon** is a group of genes with a regulatory role in protein production. The *trp* operon refers to a bacterial gene responsible for the production of the amino acid tryptophan. When tryptophan levels are high, a repressor protein binds to the operator site of the regulatory gene, switching the *trp* operon off, and resulting in no further tryptophan production. When tryptophan levels are low, the repressor protein is released and the *trp* operon is switched on. Transcription proceeds, resulting in the production of an enzyme that produces tryptophan (Figure 3.1.11).

The expression of genes can be influenced by various environmental factors, such as temperature, light and pH.

Because all cells need to engage in life-sustaining functions such as cellular respiration, the genes that control these processes are switched on in all cells. Such genes are referred to as 'housekeeping' genes.

PROTEIN STRUCTURE

Once polypeptide production is complete, the final formation of the protein can occur (Table 3.1.6).

Table 3.1.6 Levels of protein structure

1	polypeptide formation	primary protein structure
2	polypeptide becomes coiled or pleated	secondary structure
3	coiled polypeptide folds into three-dimensional form	tertiary structure
4	two or more three-dimensional polypeptide molecules bond together	quaternary structure

Proteins are large biomolecules that can contain thousands of amino acids and may be synthesised as one or several polypeptide chains. These polypeptide chains are folded and organised into specific shapes that are vital to the correct functioning of the protein (Figure 3.1.12).

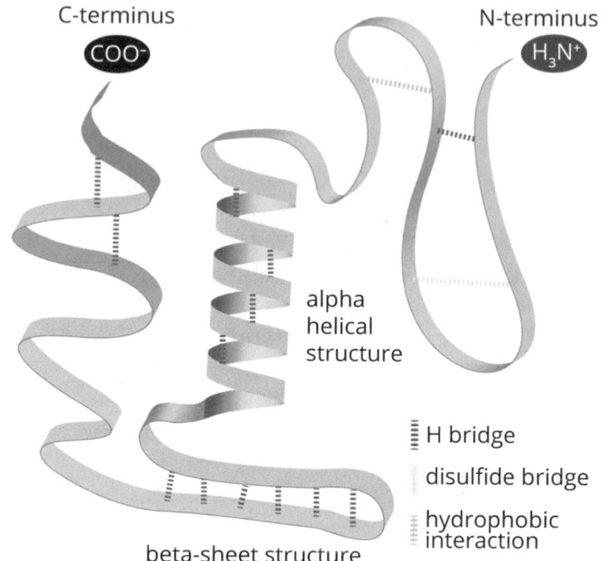

Figure 3.1.12 A simplified ribbon diagram showing segments of the primary or linear structure of a protein (polypeptide) folded into a beta sheet and alpha coil (secondary structures). The three-dimensional folding of the molecule represents the tertiary structure of the protein.

Proteins are key components of cells. There are many different kinds of proteins, each with a different function. For example, proteins in collagen have a structural role, while those embedded as plasma membrane channels have a transport role. Enzymes are of particular significance as they are involved in catalysing biochemical pathways that keep cells, and hence the entire organism, functioning. A multitude of enzymes are involved in the processes of cellular respiration and photosynthesis, as well as many other processes. Enzymes as catalysts are more fully addressed in Unit 3 Area of Study 2. Proteins in all their varied forms are vital to the normal functioning of organisms. Figure 3.1.13 illustrates some examples of the diverse kinds of proteins in living things and their roles.

ISBN 978 0 6557 0026 5

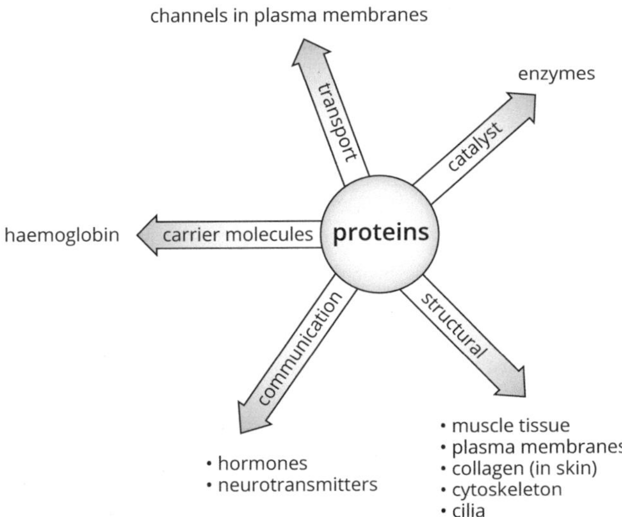

Figure 3.1.13 Proteins have many roles in living organisms.

PROTEIN SECRETORY PATHWAY

Secretory proteins are proteins that are produced to be exported from a cell. The movement of secretory proteins occurs by **exocytosis**, also known as the **protein secretory pathway**. Before reaching the plasma membrane for exocytosis, secretory proteins must first be synthesised and modified.

Proteins destined for use within the cell are synthesised by free ribosomes that are found in the cytosol. Proteins that are to be secreted are synthesised by ribosomes that stud the outer surface of the **rough endoplasmic reticulum**.

When a section of the plasma membrane wraps around a substance for import into the cell, pinching off to form a vesicle inside the cytoplasm, the process is called **endocytosis**. Pinocytosis refers to a similar process related to the import of liquid droplets. Exocytosis is the opposite of endocytosis and involves **vesicles**, such as those associated with the **Golgi apparatus**, merging with the cell's plasma membrane to facilitate the export of substances such as proteins, for example, hormones.

THE PROTEOME

The proteome is the total complement of all of the proteins in an individual organism. An organism's proteome is determined by the DNA sequence of its genome. **Proteomics** (the study of proteins including their structure and function) is an expanding field of biology that has enormous potential for increasing our understanding of how organisms function, of diseases and their treatment and management, for the development of pharmaceuticals, and for shedding light on evolutionary relationships.

Bioinformatics (the use of computers and databases to manage biological information) is a vital tool in collecting and analysing biological information, as well as making data accessible and manageable. For example, generating the DNA sequence of the human genome and making the data accessible are results of this technology.

DNA manipulation techniques and applications

Humans impact biological processes, including evolutionary processes, in all sorts of ways from selectively breeding plants and animals to serve our needs, to applying cutting-edge technology to screen and genetically modify organisms. Such interventions make humans unique in their ability to alter the course of evolutionary change in their own and other species.

GENETIC TOOLS AND TECHNOLOGIES

Advances in genetic research have made a range of tools and techniques available for DNA manipulation. Such advances have important applications in:

- medicine—diagnosis of disease and development of pharmaceuticals
- forensics—drawing conclusions related to crime investigations from DNA analysis
- evolutionary biology—DNA analysis provides information about relationships between different kinds of organisms
- bioinformatics—gathering and manipulating biological data.

Genetic tools

- Gene probe: single-stranded DNA (or RNA) sequence that is complementary to a part of the target DNA sequence used to identify the location of a particular gene or DNA fragment. Gene probes are tagged (either radioactively or fluorescently) to make them easily identifiable after binding to a target DNA sequence.
- **Primer**: short, single-stranded sequence of DNA (or RNA) that is complementary to part of the target DNA. It binds to a section of DNA that has been targeted for amplification in polymerase chain reaction (PCR), thereby identifying target DNA.
- **DNA ligase**: enzyme that joins 'sticky ends' of cut DNA strands according to complementary base-pairing rules.
- **Reverse transcriptase**: enzyme used to build 'copy DNA' (cDNA) from mRNA template, that is, transcription occurs in reverse.

KEY KNOWLEDGE

- **Restriction enzyme**: enzyme specialised for 'cutting' a DNA strand. Restriction enzymes feature a specific nucleotide sequence that is used to identify and isolate a target piece of DNA. This recognition sequence allows the restriction enzyme to bind to a 'recognition site' and sever ('cut') the bond at a particular site in the DNA sequence (sometimes called 'molecular scissors').
  - If the restriction enzyme 'cuts' the two strands in the DNA molecule in precisely the same position, the ends are called 'blunt' (Figure 3.1.14). When the restriction enzyme cuts the two strands in different positions, bases on both strands become exposed and can bind to available complementary bases. These are called 'sticky ends' because they can rejoin (Figure 3.1.14).
  - Restriction enzymes are isolated from bacteria.
  - Restriction enzymes are also called **endonucleases**.

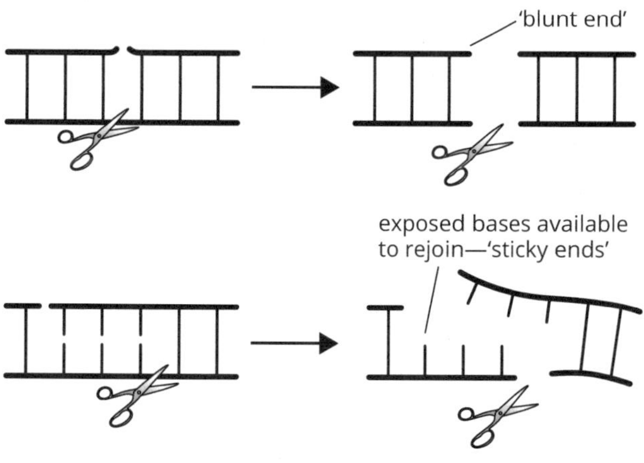

Figure 3.1.14 Restriction enzyme cuts DNA

Genetic technologies

Genetic modification describes the transfer of genes from one organism into the genome of another organism, usually of a different species. This means the organism has been genetically transformed, hence the term **genetic transformation**. Organisms that have had their genetic material modified are called **genetically modified organisms (GMOs)**. Organisms that have been genetically modified so that a gene or genes from another species have been inserted into their DNA are called **transgenic organisms**. Bacterial **plasmids** are increasingly used as vectors to confer desirable characteristics into crop plants, for example, to increase crop yield and/or enhance resistance to pests and disease. Canola, cotton and safflower represent crops subjected to these technologies in Australia.

CRISPR-Cas9 is a bacterial enzyme complex that allows for genome editing in living cells. CRISPR is an abbreviation for 'clustered regularly interspaced short palindromic repeats'. Essentially, short repetitive DNA sequences that form palindromes (read the same both forwards and backwards) occur in clusters with regular spacing between them. The spacers contain unique DNA segments (that is, not repeats). Scientists believe the unique DNA sequences are copies of viral DNA retained by the bacteria after infection and represent a kind of memory, allowing the bacteria to quickly produce RNA that is complementary to the CRISPR arrays (known as CRISPR RNA or crRNA). The bacteria use the crRNA to target the virus' DNA and cut it with endonuclease enzymes, disabling the virus. The enzyme that cuts the viral DNA is a CRISPR associated protein called Cas9. The CRISPR-Cas9 system allows the bacteria to combat subsequent exposures to the same virus more effectively. Scientists have developed a CRISPR-Cas9 system that works in a similar way in laboratory conditions, enabling them to edit DNA sequences in living cells. This genetic tool has far-reaching potential for altering the genomes of organisms to confer the most desirable characteristics, including crops and livestock as well as correcting faulty genes that are the cause of genetic disorders.

Gel electrophoresis compares the distance that specific DNA fragments travel along a gel preparation that is subject to an electric current (to drive the DNA through the gel) (Figure 3.1.15). Smaller DNA fragments move further through the gel compared to larger fragments. The DNA samples being analysed are compared to standards (DNA fragments of known length) to determine the length of the sample DNA. Length of DNA fragments are measured in 'base pairs' (bp).

DNA profiling, also called DNA fingerprinting, uses a pattern of repeated DNA sequences (called 'short tandem repeats'—STRs) that are unique to an individual to identify a particular person's DNA (Figure 3.1.16). The DNA profile is observed using gel electrophoresis technology. DNA profiling is used to establish relationships between individuals, for example, to determine paternity or to confirm or exonerate crime suspects.

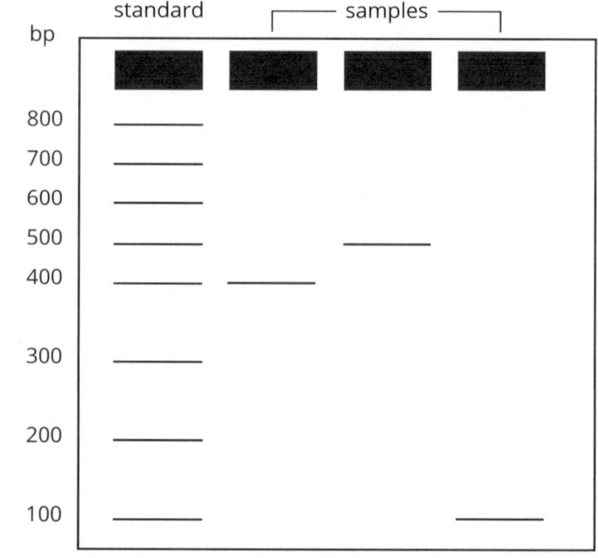

Figure 3.1.15 Gel electrophoresis

 ISBN 978 0 6557 0026 5

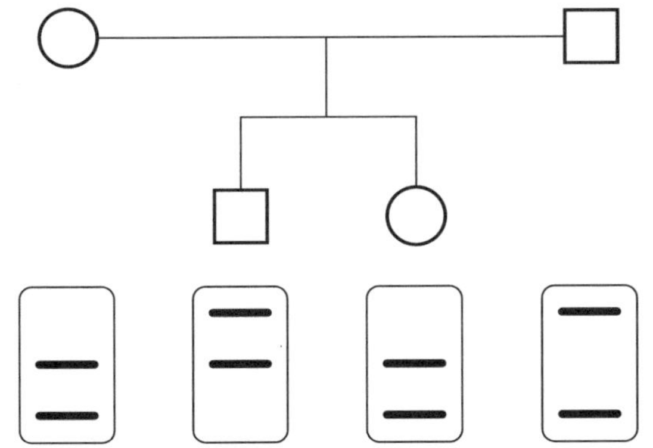

Figure 3.1.16 Family relationships evident using DNA profiling

Polymerase chain reaction (PCR) is a technique that amplifies or copies a target fragment of DNA. PCR makes large volumes of target DNA available for other applications, for example, gel electrophoresis and DNA profiling (Figure 3.1.17).

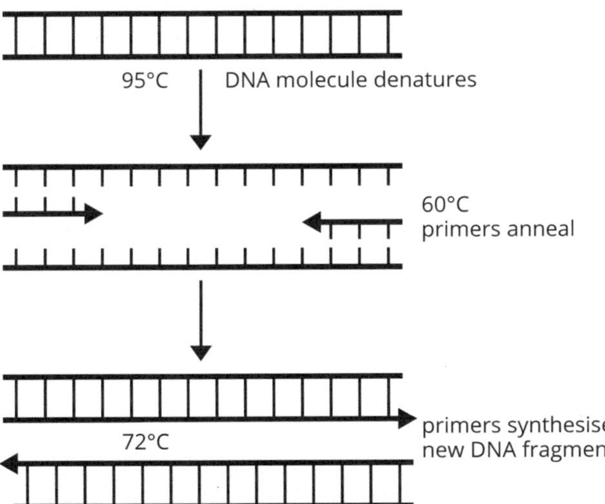

Figure 3.1.17 PCR amplifies a DNA fragment.

DNA sequencing is a process that is used to determine the order of nucleotide bases along a segment of DNA. Bases are 'tagged' so that each appears a different colour when viewed under fluorescent light. Chromatography is used to observe the tagged bases in a series of coloured peaks. The order of the coloured peaks reflects the order of the bases in the DNA strand (Figure 3.1.18). Identifying the base sequence of the human genome in the Human Genome Project is a classic example of the use of this technology. This technology is also used to compare DNA sequences from different species, establishing how closely related they are.

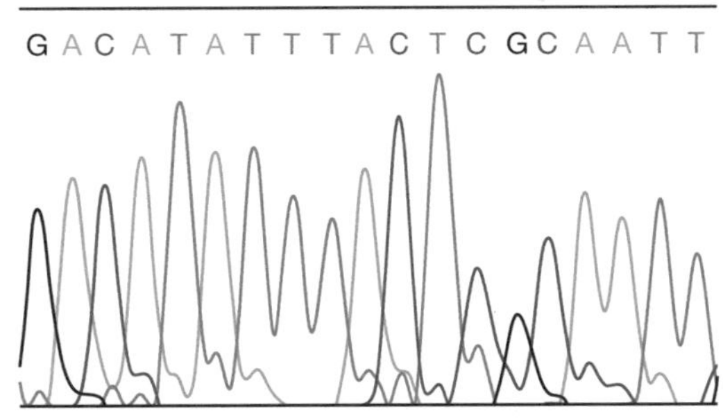

Figure 3.1.18 DNA sequencing. Each peak represents a nucleotide.

Gene cloning involves making copies of a selected gene. This technology is applied in the genetic modification of bacterial DNA to construct **recombinant plasmids**. Such transgenic bacteria are important in the genetic modification of crop plants such as Bt cotton. Gene cloning also has medical applications, for example, in the production of human insulin for the treatment of diabetes.

Gene therapy is a process that imports a healthy gene into the DNA of a vector, such as a disabled virus, to treat diseases caused by the inheritance of defective genes. Cystic fibrosis is an example of such a genetic condition.

Cloning technology is also applied to create genetically identical individuals. Cloning of plants is easy to do from cuttings and has been a long-standing practice in agriculture. Animal cloning involves removing a diploid nucleus from a somatic cell and inserting it into an emptied ovum that is ready for fertilisation. After a period of laboratory incubation, the developing embryo is implanted into the uterus of an adult female. A normal pregnancy ensues and the resulting offspring is a genetic clone of the donor of the somatic cell nucleus. Cloning is used in the production of crops and livestock with desirable characteristics, for example, pest-resistant crops that increase overall crop yield. Therapeutic cloning is used to produce tissue that is suitable for transplant in humans, for example, to produce new skin for burns victims.

Genetic screening is a technique used to identify the presence or absence of genetic disorders, which may be caused by defective genes or abnormal chromosome numbers. This procedure is an important tool in assessing both the chance of developing the genetic disease in question or passing it on to potential offspring. As such, it provides information to assist decision-making.

WORKSHEET 1

Knowledge review—scientific method and genetic technologies

Scientific method

The scientific method is a vital tool that ensures a sound approach to investigations that yield reliable data and logical conclusions. By following the scientific method, researchers can contribute to the development of rigorous biological principles.

1 A student decided to test the idea that potatoes left in a dark cupboard can sprout stalks by vegetative reproduction. To test this idea the student placed 10 similar sized potatoes in a dark cupboard and another 10 potatoes on the kitchen bench for four weeks.

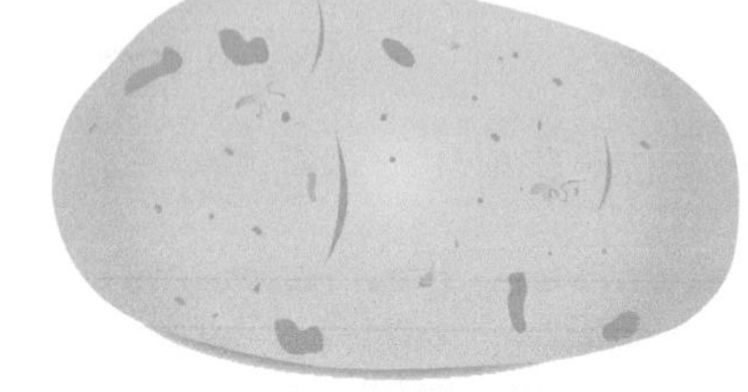

a Complete the table by entering the definition for each term and the instance of each in this experiment.

Element of experiment	Definition	In this experiment
hypothesis		
independent variable		
dependent variable		

At the end of the test period the student observed that all of the potatoes in the dark cupboard had grown stalks, some short and some long, ranging from 2 to 6 cm in length. The potatoes placed on the kitchen bench also showed some growth, but the stalks were much shorter and there were fewer of them.

b Clarify which of the student's observations represent qualitative data and which represent quantitative data. Explain the difference.

 ISBN 978 0 6557 0026 5

Genetic technologies

This part of the activity aims to remind you of some genetic technologies you have heard about before and to consider your understanding and perspective about them.

2 **a** Consider each of the genetic technologies listed in the table below. Prepare a description outlining your understanding of the technology and write one or two questions the technology raises.

Genetic technology	Description	Questions
GM foods		
DNA profiling		
cloning		

b Discuss your responses with other members of the class. Add to your own list. Use online resources to refine your understanding in each case.

3 Scientific advances in genetic research have made technologies available that are important in medicine, forensics and evolutionary biology. Such technological advancements deliver advantageous outcomes for many, but also raise issues that are often the subject of debate in the community.

Select one genetic technology you have heard about and outline two benefits and two concerns.

Genetic technology: ______________________________

a benefits

____

____

b concerns

____

____

ISBN 978 0 6557 0026 5

WORKSHEET 2

Classification and identification • Modelling

The DNA molecule

1 **a** Name the molecule below. ____________________

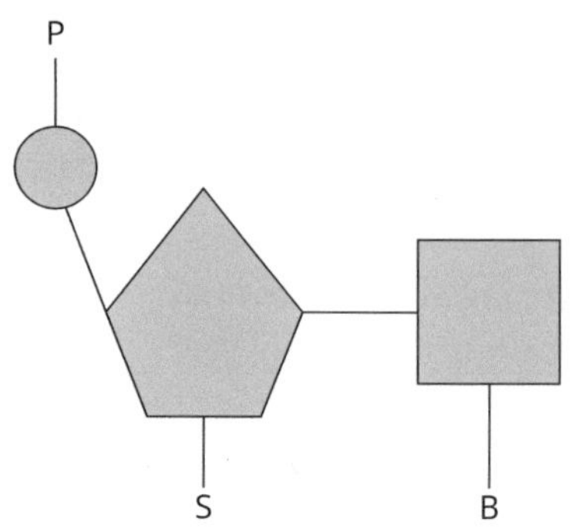

b Identify the parts of the molecule labelled:

P ____________________

S ____________________

B ____________________

2 Examine the section of a DNA molecule below.

5'

C

A

G

C

T

C

A

C

T

3'

a What do the letters A, T, C and G represent?

____

b Describe the relationship between these units.

____

c Draw in the complementary DNA strand.

d Explain the term 'antiparallel' as it relates to the DNA molecule.

____

 ISBN 978 0 6557 0026 5

WORKSHEET 3

Genes, genome and the role of DNA

An organism's genome and its corresponding proteome can be likened to a dictionary in which the genome represents all of the words and the proteome outlines the meanings of those words. The genome includes all of the genes that are the instructions for building all of the proteins in the body.

In this concept map you begin with the genome and follow the steps to finally reach the cell's functional destination—its proteome.

Use the boxes to write the definitions of the terms. Add your own branches and words to the concept map to build a word picture of the key structures and processes that occur as the cell follows its genetic instructions to build its proteins.

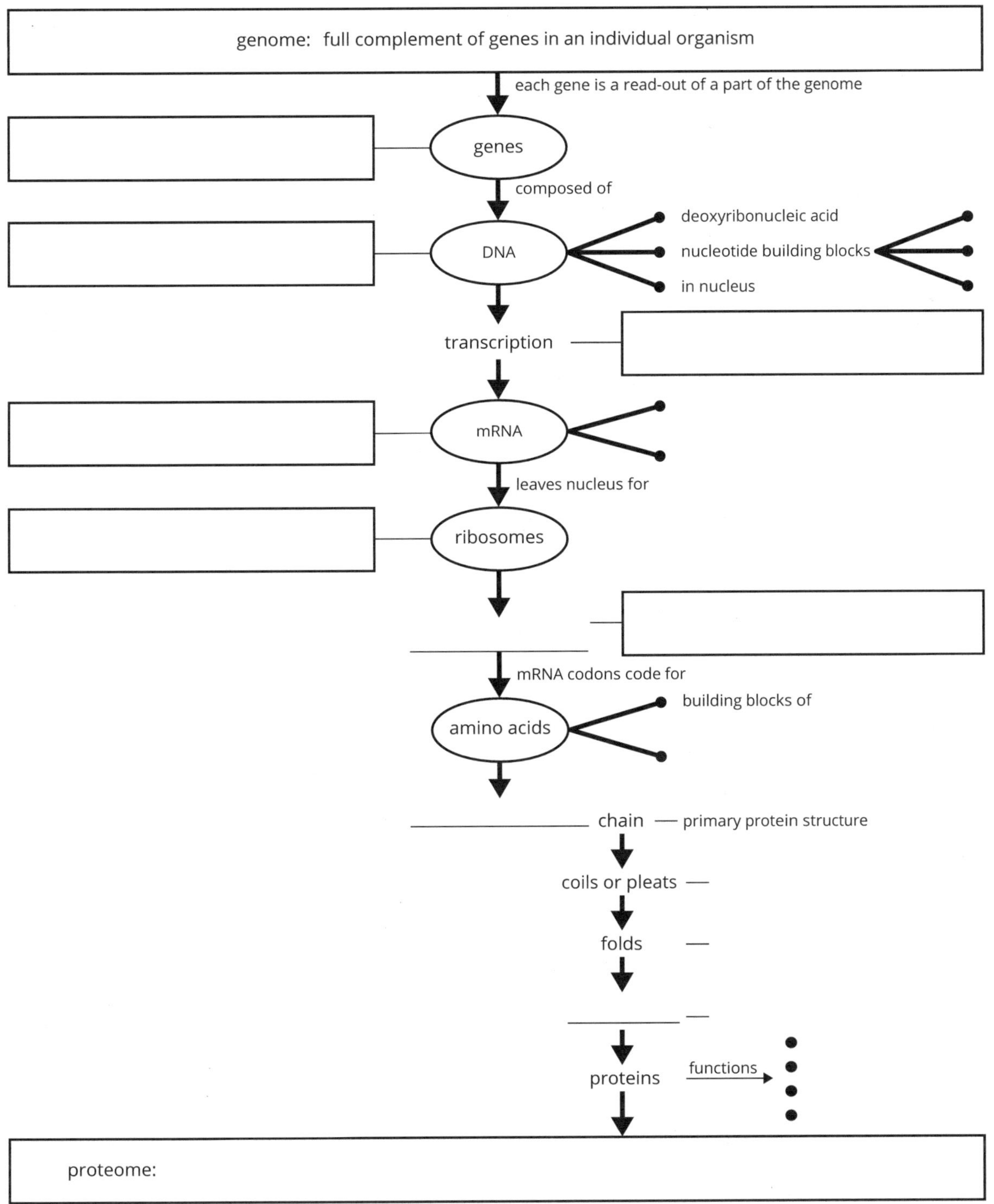

WORKSHEET 4

Classification and identification • Modelling

Transcription and translation

DNA is a biological book of words and meanings that can be likened to a dictionary where:

- the words represent the genes
- the definitions are an expression of what each word means. In this case, the expression of a gene is the protein for which it codes.

The figure below shows each step involved in reading the DNA sequence of a gene segment. By following each step, the final meaning is revealed.

1 Use your knowledge of complementary base-pairing rules and the genetic code to add in the missing instructions in the diagram below.

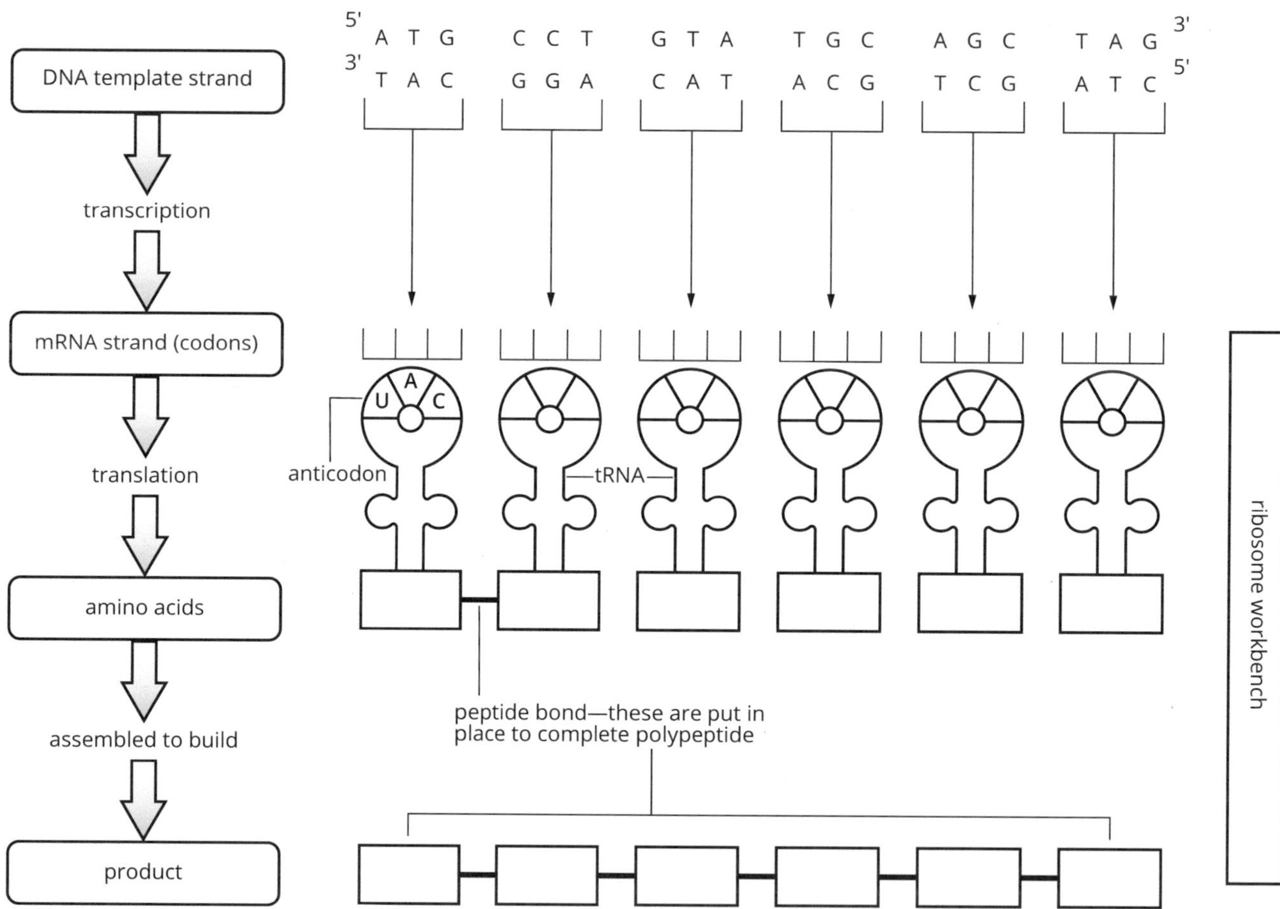

2 Describe what happens during:

a transcription

____

____

____

b translation

____

____

____

 ISBN 978 0 6557 0026 5

3 Name the first and last codons in the above sequence. Outline their significance.

4 Identify the name of the product that has been constructed at the end of the process.

5 Use a single sentence to outline the fundamental role of genes.

WORKSHEET 5

Gene regulation and the *trp* operon

An operon is a group of genes with a regulatory role in protein production. Francois Jacob and Jacques Lucien Monod were awarded the Nobel Prize in 1965 for their pioneering work in discovering the operon and its role in regulating gene expression. Their research centred on genes involved in the production of the amino acid tryptophan in the bacterium *E. coli*. In 1953, Monod and his colleagues first identified a regulatory system of genes. This gene regulatory system was called the tryptophan operon, more commonly referred to as the *trp* operon.

An operon is composed of a regulatory gene, typically containing a promoter region and an operator site, and the structural gene. Together this DNA unit is transcribed into a segment of mRNA. The mRNA codes for a protein. In the *trp* operon, the structural gene contains a series of genes that code for the production of three enzymes involved in the tryptophan synthesis pathway. In the regulation of tryptophan production, the repressor protein gene is separated from the rest of the regulatory system along the bacterial DNA.

Monod and his team discovered that the mechanism for regulating tryptophan production involves a repressor protein binding to the operator site of the regulatory gene. When tryptophan concentrations in the cytoplasm are high, molecules of tryptophan bind with repressor proteins, which in turn attach to the operator, blocking transcription (essentially switching the gene off). As tryptophan production ceases, low tryptophan levels trigger the release of the repressor protein from the operator, switching the gene back on so that transcription resumes, along with tryptophan synthesis.

1 Select from the terms below to correctly label the *trp* operon.

tryptophan enzyme gene/s	operator region	repressor protein gene

trp operon

regulatory gene | structural genes

		promoter region		*trpE*	*trpD*	*trpC*	*trpB*	*trpA*	stop

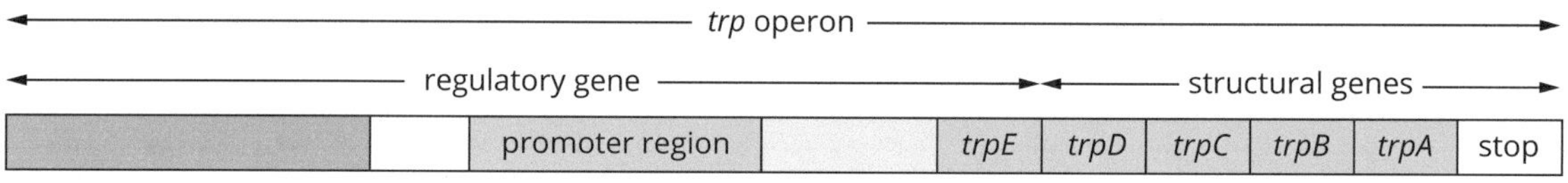

a ____________________

b ____________________

c ____________________

2 Outline the difference between the regulatory region and the structural gene in an operon.

____

____

____

ISBN 978 0 6557 0026 5

WORKSHEET 5

Figure 3.1.19 illustrates the sequence of events in the regulation of the tryptophan gene and its expression.

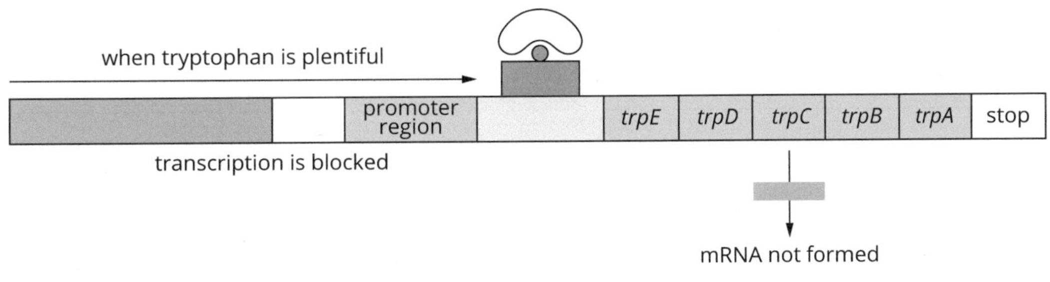

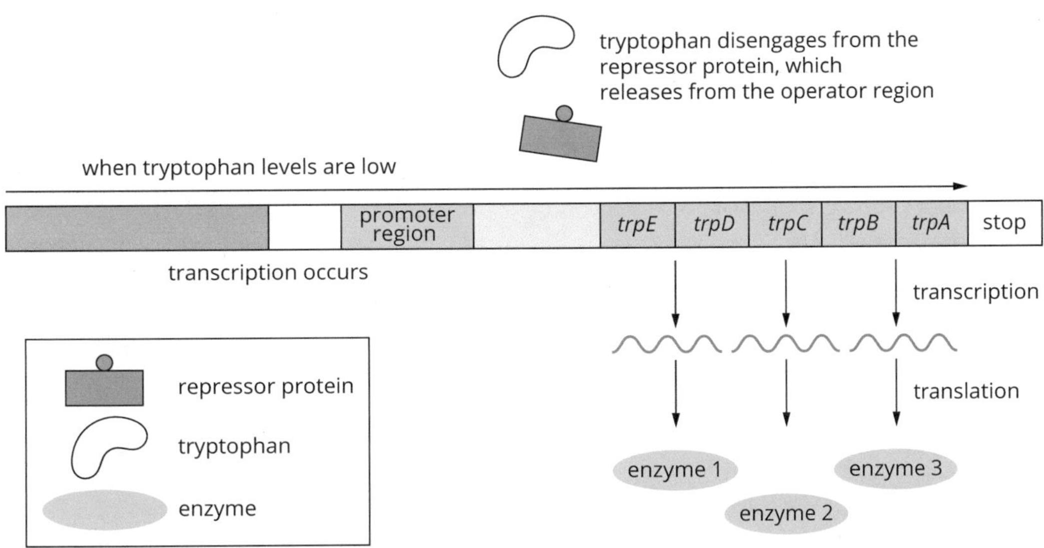

Figure 3.1.19 Regulation in the *trp* operon

3 **a** Identify the protein product of translation of structural genes in the *trp* operon.

b Describe the role of this product.

4 Suggest why tryptophan is of value to *E. coli* bacteria.

5 What is the advantage to the *E. coli* of blocking transcription of the tryptophan gene when the amino acid is plentiful?

6 Summarise the role and significance of regulatory mechanisms in the expression of genes for organisms.

WORKSHEET 6

Genetic tools and technologies

1 Make a selection from the list below to fill in the missing words in each summary statement.

Taq polymerase	cloning	recombinant DNA	primers	gel electrophoresis
DNA ligase	endonucleases	DNA probes	genetic modification	DNA profiling
polymerase chain reaction	genome	DNA sequencing	plasmid	

- Genetic tools and techniques are used to investigate the genomes of different species, to isolate DNA sequences for medical research and to solve crimes, as well as to produce genetically modified organisms (GMOs).
- The biological technology that allows genes from one organism to be transferred into another is called ______________ ______________. This process is common in commercial crop plants today because it confers desirable characteristics, such as resistance to particular insect pests. Such a feature can reduce costs because a smaller quantity of insecticide is needed; it is important for health reasons because fewer residues remain on crop products bound for the marketplace.
- Restriction enzymes, also called ______________, take on the role of 'molecular scissors', allowing geneticists to 'cut out' or isolate specific fragments of DNA for analysis or amplification.
- Once isolated, specific DNA fragments can be amplified or copied in large numbers using the process of ______________ ______________ ______________. This process involves the separation of a double-stranded DNA fragment at high temperatures and the subsequent replication of the original DNA. The enzyme _________ ______________ puts nucleotides in place according to base-pairing rules.
- The size (in base pairs) of a particular DNA fragment can be assessed by measuring the distance it moves along a gel plate driven by an electric current compared to a known DNA standard fragment, in a process called _________ ______________.
- _________ ______________ is a technology used to identify the order of nucleotide bases in a strand of DNA.
- Radioactive isotopes and fluorescent dyes used to label specific segments of DNA are called ______________ ______________.
- Genetic technologies are important in medical research. For example, people who have diseases such as diabetes are able to reap the benefits of gene cloning—a process in which _________ ______________ technology introduces the normal human insulin-producing gene into bacterial DNA, resulting in ample production of human insulin to treat patients.

2 You will see that there are six terms remaining. Define each of these terms.

____

____

____

____

____

____

 ISBN 978 0 6557 0026 5

WORKSHEET 7

Case study

Gel electrophoresis in species conservation

Figure 3.1.20 The Amur leopard (*Panthera pardus orientalis*)

Gel electrophoresis is a critical tool for analysing genetic relationships in a range of applications, including familial relationships, forensic investigation and evolutionary relationships. It also has an important role in the conservation of species, particularly Critically Endangered species such as the Amur leopard, *Panthera pardus orientalis* (Figure 3.1.20). Dwindling wild populations of the leopard continue to be threatened by destruction of their habitat to make way for human population growth, and illegal poaching for their coats (Figure 3.1.21). With as few as 70 Amur leopards estimated to be living in the wild, breeding programs between zoos are vital in keeping them from extinction. Gel electrophoresis allows zoos to identify the closeness of the relationship between a potential breeding pair.

The gel electrophoresis shown in Figure 3.1.22 considers several DNA markers for a female Amur leopard and three potential mates from other zoos. Examine the completed gel run. Use the information to answer the questions.

Figure 3.1.21 Black indicates the only remaining wild populations of the Amur leopard.

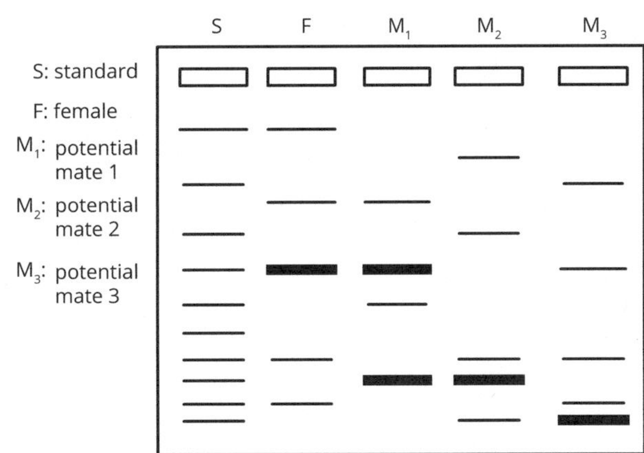

Figure 3.1.22 Gel electrophoresis

1 Identify the pair of Amur leopards you would recommend for breeding. Explain your choice.

2 One characteristic of many endangered species is small population size. What impact might this have on the gene pool of a species?

3 To what extent is a technique such as gel electrophoresis useful in captive breeding programs?

WORKSHEET 8

Literature review • Modelling

Genetically modified bacteria and insulin production

Plasmids are loops of DNA present in the cells of bacteria. Plasmids are typically composed of a relatively small number of base pairs and contain few genes. They are able to reproduce independently of the bacterial chromosome. Plasmids are easily cut and readily accept foreign DNA; the exposed DNA ends bind easily to once again close the loop. These features make plasmids especially useful as vectors in transforming bacterial cells. The transformed bacteria can be cultured, with the recombinant plasmids continuing to replicate. The foreign genes are transcribed and translated within the bacteria to synthesise large quantities of the desired protein.

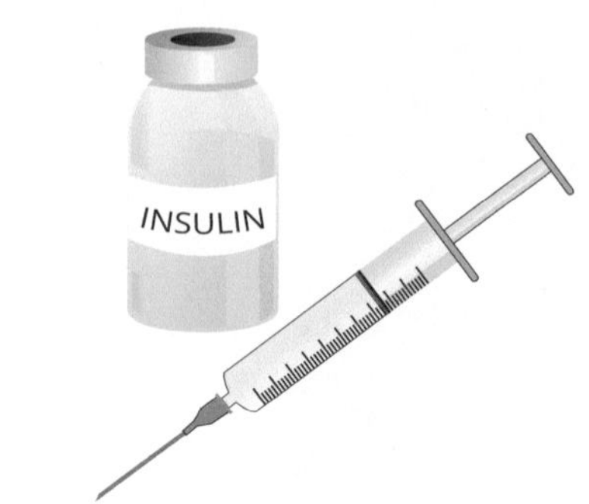

Figure 3.1.23 Insulin is used to treat diabetes.

In this activity you will research online for information illustrating the steps involved in the large-scale production of human insulin (Figures 3.1.23 and 3.1.24), which includes the use of genetically modified bacteria.

1 The production of insulin to treat diabetes is one significant application of the use of recombinant plasmids.

   Research online for information on how insulin is made, including the steps involved in modifying bacterial plasmids to manufacture insulin. An animation will be helpful in illustrating the process.

   In the space below, prepare a flow chart of the steps in bacterial transformation in the manufacture of insulin. Include relevant terminology and diagrams to illustrate key steps in the process.

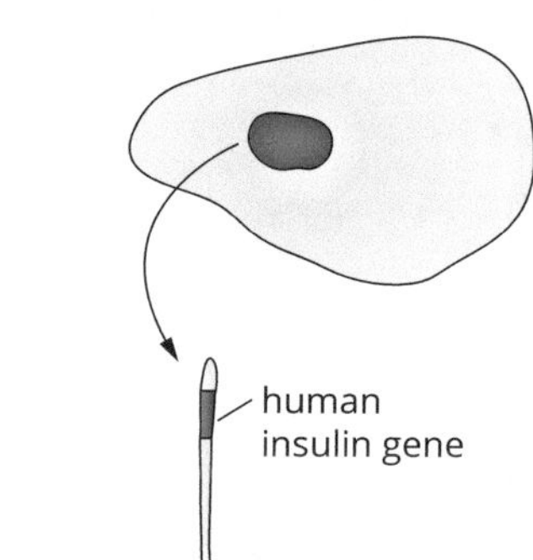

Figure 3.1.24 Human insulin gene identified for isolation

ISBN 978 0 6557 0026 5

WORKSHEET 8

2 How can geneticists be certain about which plasmids have taken up the insulin gene and which have not?

3 Why is it important to identify with certainty the transformed bacteria?

4 Outline an advantage of using transformed bacteria as a source of insulin rather than the more traditional approach of using pig insulin.

WORKSHEET 9

Case study • Modelling

Applications of genetic transformation

The year 2003 saw a novel direction in human intervention in the natural world. A new breed of designer pet appeared for sale in some countries—a glow-in-the-dark, genetically modified zebrafish (Figure 3.1.25).

The curious ornamental fish has been the subject of gene technologies that allow the gene in question to be isolated from sea jellies and then inserted into fish embryos at a very early stage of development—the one- or two-cell stage.

Figure 3.1.25 Transgenic fluorescent zebrafish (*Danio rerio*)

1 Use your knowledge of gene technologies to draw a flow chart of the steps that occur in genetic transformations of this kind. Include genetic tools used in the technology. Diagrams might also be helpful.

ISBN 978 0 6557 0026 5

2 Suggest the likely fate of a transgenic fluorescent zebrafish if it were released into wild populations. Explain your answer.

3 The fluorescent zebrafish has been the subject of heated debate both in Australia and overseas, and has been banned from sale in some countries. Form a group to discuss the ethical issues raised by this technology. Prepare a list of issues.

4 As genetic technologies become more refined they are applied to increasing numbers of organisms, often for commercial means. Canola is one such crop to which genetic technology has been applied in Australia (Figure 3.1.26).

Enter keywords into a search engine to find out more about the genetic transformation of canola.

a Suggest a reason that canola producers may favour genetically modified canola.

Figure 3.1.26 Canola crop

b Suggest a reason why consumers may be concerned about genetically modified food crops.

5 Write your own personal response to the application of genetic technology in general, and the fluorescent zebrafish in particular.

ISBN 978 0 6557 0026 5

WORKSHEET 10

Reflection—What is the role of nucleic acids and proteins in maintaining life?

The following table lists the key knowledge covered in this area of study.

1 Reflect on how well you understand the concepts listed. Rate your learning by shading the circle that corresponds to your current level of understanding for each one.

Key knowledge	Not confident ◄			► Very confident
Structure and function of DNA and RNA	○	○	○	○
Gene expression, including transcription, RNA processing and translation	○	○	○	○
Gene structure: exons, introns and promoter and operator regions	○	○	○	○
Gene regulation and the *trp* operon	○	○	○	○
Structure and function of proteins	○	○	○	○
The protein secretory pathway and the role of the rough endoplasmic reticulum, Golgi apparatus and vesicles	○	○	○	○
Enzymes that are used to manipulate DNA, including polymerase and ligase	○	○	○	○
CRISPR-Cas9 and its use in genome editing	○	○	○	○
Polymerase chain reaction and gel electrophoresis	○	○	○	○
Recombinant plasmids as vectors	○	○	○	○
Genetically modified and transgenic organisms	○	○	○	○

2 Consider the points you have shaded from Not confident to Very confident. List specific ideas you can identify that were challenging.

3 Write down two different strategies that you will apply to help further your understanding of these ideas.

ISBN 978 0 6557 0026 5

PRACTICAL ACTIVITY 1

Case study • Classification and identification

An examination of processes and issues around genetically modified food crops

Suggested duration: 60 minutes

INTRODUCTION

The development of genetically modified (GM) organisms for human consumption is increasing, both in Australia and overseas. GM crops approved for growing in Australia are currently limited to canola, cotton and safflower, however, many more are imported and used in processed foods (Figure 3.1.27). GM foods imported into Australia and used as ingredients in the processing of other foods include potatoes, corn, soybeans, sugar beets and rice. Research trials into the development of other GM food plants for approval in Australia continue and include wheat, barley, pineapples, bananas, rice, corn and sugarcane. Genetic modification of such crop plants is intended to confer desirable qualities such as resistance to pests, increased herbicide tolerance or enhanced yield.

Figure 3.1.27 GM canola and some canola products

AIM

- To consider an application of GM technology for Australian food supply.
- To consider some advantages and disadvantages of GM foods.
- To consider some ethical issues raised by GM technology in relation to foods.

Cultivating GM canola

Commonly known as canola or rapeseed, *Brassica napus* is a widely grown crop in Australia used in the production of vegetable oils and spreads such as margarine. Canola oil is an ingredient in many dairy blends and canned foods. Canola crops are a familiar site in rural areas, identifiable by a sea of fine, yellow flowers. Unfortunately, canola crops compete with weeds for space and soil nutrients, which compromises plant quality and crop yield. The use of traditional herbicides to control weeds also impacts on canola plants. Genetically modified varieties of canola are tolerant to herbicides, that is, application of herbicides kills the weeds but does not damage the tolerant canola plants.

The soil bacterium *Agrobacterium tumefaciens* is typically used as a vector in the genetic transformation of canola. After bacterial plasmids are treated to carry a herbicide-resistant gene, the transformed bacteria are introduced to tissue culture from the canola plant. The bacteria infect the plant cells, which grow into an undifferentiated mass called a callus. Each callus grows into a plantlet that exhibits the desired herbicide resistance. The genetic transformation of *Agrobacterium tumefaciens* is shown, in part, on page 28.

PRACTICAL ACTIVITY 1

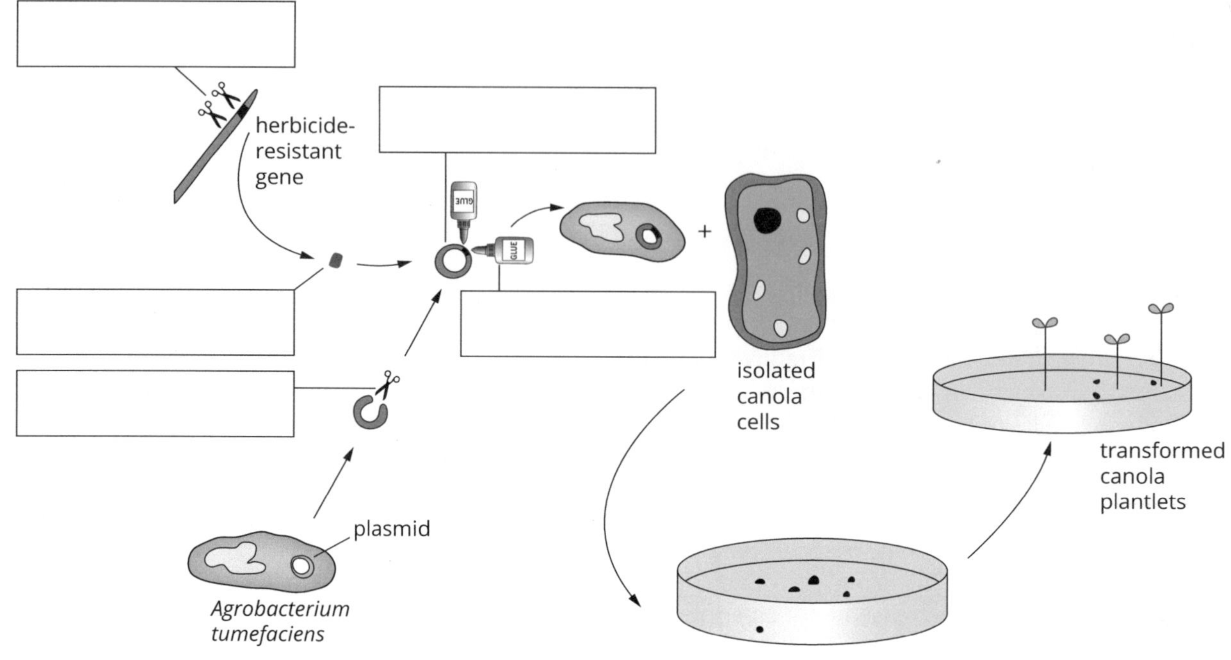

1 Select from the terms below to correctly label the stages of the GM process for canola, above. Terms may be used more than once.

endonuclease	recombinant plasmid	herbicide-resistant gene	ligase

2 Explain why *Agrobacterium tumefaciens* is described as a vector in this process.

3 Outline the role of the following elements in the GM process for canola.

a *Agrobacterium tumefaciens*

b endonuclease

c ligase

4 Explain why it is important to apply the same endonuclease to cut the desired gene from the surrounding DNA and to cut the plasmid open.

 ISBN 978 0 6557 0026 5

PRACTICAL ACTIVITY 1

5 Is the GM canola represented in the diagram best described as genetically modified or transgenic, or could both be applied? Explain.

6 Outline an advantage and a disadvantage related to GM canola.

Advantage:

Disadvantage:

Ethical considerations

There is considerable debate in the community about the development of genetically modified organisms for human consumption. In Australia, GM foods must include this status in their labelling (Figure 3.1.28). Livestock for human consumption in Australia is not genetically modified, however, these animals may be fed a diet of GM food. While the law in Australia is clear about labelling genetically modified produce as GM, there are exceptions that mean some foods containing GM products may not need to be labelled as such. Highly refined foods such as cooking oil, margarines and chocolate may fall into this category. So do bakery goods that may have some GM ingredients added to them.

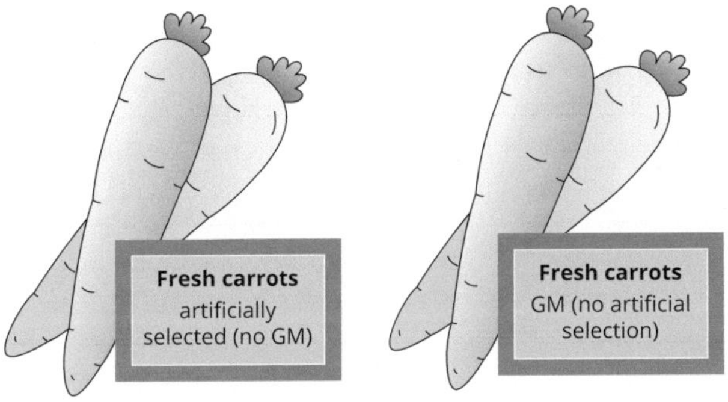

Figure 3.1.28 In Australia, GM foods must include their status in their labelling.

7 Make a list of some community concerns about GM foods.

8 Consider the following statement.

'Genetic modification of organisms to ensure the most desirable characteristics is simply taking a short cut to achieve the same outcome that hundreds or thousands of years of selective breeding can do.'

Explain whether you agree or disagree.

9 Explain whether you agree or disagree with Australia's labelling laws in relation to GM foods.

10 Suggest issues that may face farmers of crops such as canola who wish to grow and market their product as GM-free.

CONCLUSIONS

11 How are GM foods different to non-GM foods?

12 Under what circumstances are primary producers likely to win approval for growing GM crops for human consumption in Australia?

FURTHER INVESTIGATION

- Check the labelling on foodstuffs in your pantry and on supermarket shelves to identify which foods contain GM ingredients.
- Some people prefer foods to be 'natural', without any changes to their 'natural' DNA. Research just how natural the foods we eat are. For example, how are commercially grown carrots different from 'natural' or wild carrots? What about watermelon and bananas? How have we achieved the plant breeds we eat today? How is this different from GM food?

ISBN 978 0 6557 0026 5

PRACTICAL ACTIVITY 2

Case study • Modelling

A case study of Huntington's disease including gel electrophoresis and a focus on issues

Suggested duration: 60 minutes

INTRODUCTION

Gel electrophoresis is a genetic technology used to sort individual strands of DNA according to their length (Figure 3.1.29). The technique uses a gel prepared from agarose powder and buffer solution. Ethidium bromide dye, which binds to DNA making it visible under UV light, is added to the mixture. The gel is poured into a mould the shape of a small tray. At one end, a 'comb' is inserted into the gel and the gel is allowed to cool. When the comb is removed a series of 'loading wells' is left in the gel. DNA samples identified for testing are placed in some of the wells. 'DNA standards' are loaded into other wells—these represent DNA fragments of known size. Once the wells have been loaded, the gel is subjected to an electric current. Negatively charged DNA fragments are dragged through the gel towards the positive terminal. Smaller DNA fragments move further through the gel than larger fragments. At the end of the gel run, comparison of the 'finishing' positions of the various fragments is used to determine their respective lengths. For the purposes of gel electrophoresis, the length of DNA fragments is measured in 'base pairs' (bp). The size of sample DNA fragments is determined by comparison with known size standards.

AIM

- To consider a case study of a family in which there is a history of Huntington's disease.
- To use the technique of gel electrophoresis in determining the genetic status of an individual in relation to Huntington's disease.
- To explore some issues that arise from the information provided by genetic technologies.

BACKGROUND

Huntington's disease is an inherited disorder that is characterised by the progressive degeneration of neurological function. The symptoms of this disease include clumsy, involuntary movements, difficulty in swallowing and speaking, hallucinations and a decrease in mental capabilities. Eventually, the patient is completely dependent on carers. Onset of the disease usually occurs in the 30s or 40s and death generally occurs between eight and 20 years after onset. The mode of inheritance for Huntington's disease is autosomal dominant.

Huntington's disease is caused by a change in the base sequence of the *HD* gene, which controls production of the IT15 protein. This protein is important for normal neurological function. The normal protein is composed of a polypeptide chain containing a sequence of 3144 amino acids. The coding region of the *HD* gene has a base pair length of 9432. The normal nucleotide sequence of the gene contains between 15 and 30 repeats of the base triplet GTC (mRNA sequence CAG), which codes for the amino acid glutamine. Huntington's disease is called a trinucleotide repeat disease because there are many more repeats of this triplet than exist in the normal gene. The added amino acids affect the structure and function of the IT15 protein.

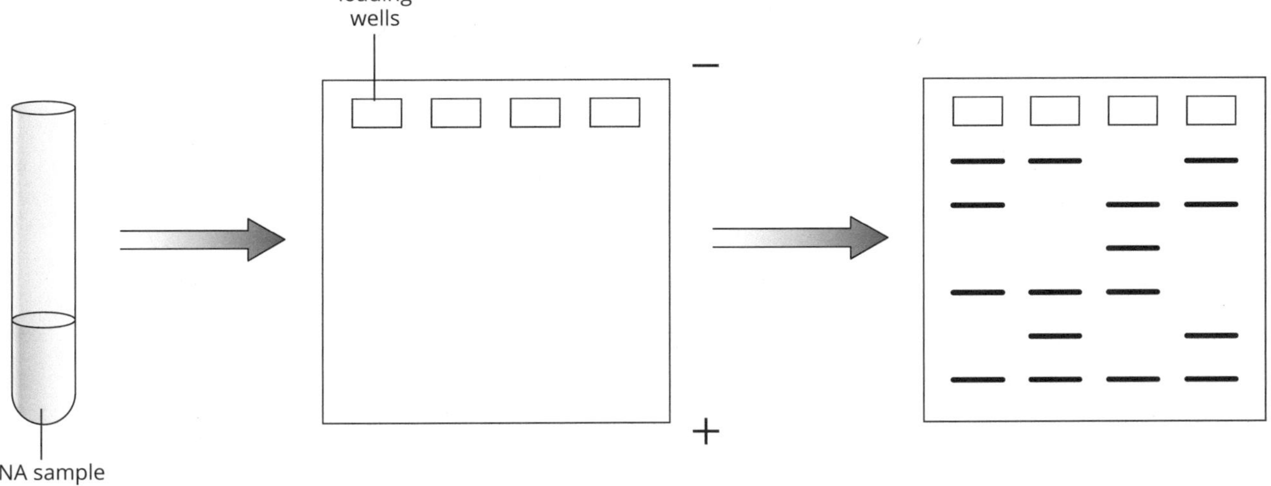

Figure 3.1.29 Gel electrophoresis

PRACTICAL ACTIVITY 2

Case study

Carl, a 25-year-old man whose mother suffers from Huntington's disease, is planning a family with his partner but has concerns about the possibility of passing on Huntington's disease to any children he may have. He understands that the disease is inherited, but he has shown no signs of the disease himself. He attends a genetic diseases clinic where the technique of gel electrophoresis is used to assess his genetic status in relation to this disease. Other members of his family are also tested.

A DNA sample is taken and the relevant DNA fragment is identified, and then amplified using PCR. Abnormalities in the number of triplet repeats occur in the first 200 base pair sequence of the *HD* gene. The results of Carl's gel run are shown in Figure 3.1.30.

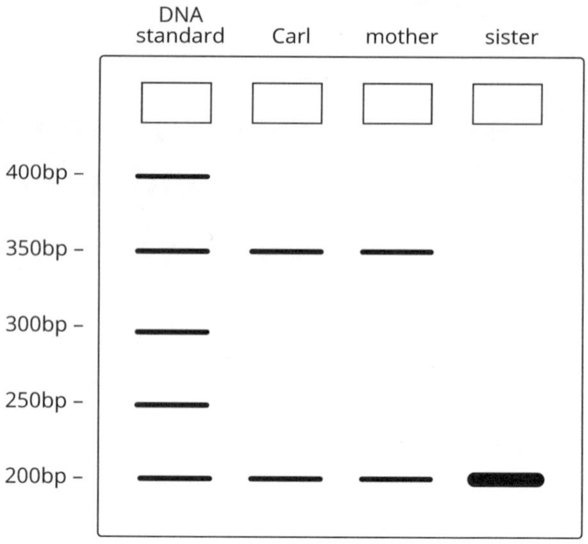

Figure 3.1.30 Gel electrophoresis data

1 How is gel electrophoresis useful to geneticists?

2 Describe the factor that determines how far the different strands of DNA move through the gel.

3 Outline the measures taken to ensure the DNA moving through the gel is visible.

4 Examine the gel electrophoresis results for Carl in Figure 3.1.30.

a Why are there two bands?

b What is the size of the DNA fragments?

c What does this mean in terms of the number of triplet repeats of GTC in the *HD* gene in Carl's DNA?

ISBN 978 0 6557 0026 5

d How does this compare with the normal gene?

5 a What does the gel run reveal about Carl's status in relation to Huntington's disease?

b Carl's partner has no history of Huntington's disease in her family. What are the chances of any of their children having Huntington's disease? Use notation and a Punnett square to explain your answer.

6 Make a list of the issues faced by Carl and his partner in light of the test results.

7 Carl's sister was also tested. Describe her results and explain what they mean for her and any children she may have.

PRACTICAL ACTIVITY 2

SUMMARY REPORT

8 Prepare a report summarising the process of gel electrophoresis and its importance in modern science. Make sure your report includes:

- a summary statement describing how gel electrophoresis is used to manipulate DNA
- a flow chart of the key points in the process of gel electrophoresis
- an outline of how gel electrophoresis is useful in diagnosing inherited disorders
- a discussion of the ethical issues raised by the development and use of genetic technologies.

ISBN 978 0 6557 0026 5

PRACTICAL ACTIVITY 3

Simulation

A PCR simulation

Suggested duration: 60 minutes

INTRODUCTION

A significant quantity of DNA is needed for many DNA manipulations. When only small or trace amounts are available, as may occur, for example, in a crime scene, polymerase chain reaction (PCR) can be applied to amplify that DNA. This makes millions of identical copies of the target DNA available for other applications such as gel electrophoresis and DNA profiling (Figure 3.1.31).

This investigation involves participating in a virtual laboratory activity. You will need to search the internet for a virtual PCR laboratory, such as the one at the DNA Learning Center at Cold Spring Harbor Laboratory, USA.

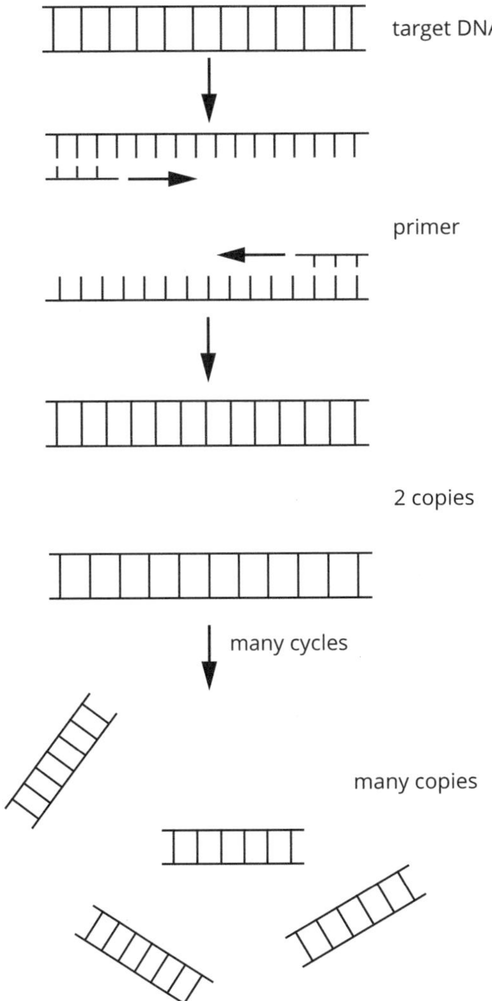

Figure 3.1.31 The polymerase chain reaction (PCR) process

AIM

- To simulate the process and outcome of polymerase chain reaction.
- To consider applications of polymerase chain reaction.

METHOD

Use key words to search for a virtual PCR laboratory. As a guide, the DNA Learning Center at Cold Spring Harbor Laboratory, USA provides a suitable example. Enter the site and read the information before launching the virtual laboratory. Follow the prompts to complete each step of the virtual PCR activity, answering the questions below as you proceed.

1 Before undertaking PCR, DNA needs to be extracted from the nucleus of cells. List three possible sources of cells that could provide DNA for PCR.

2 What are primers? Describe their role in this reaction.

3 Name the four different nucleotide bases that have been added to the buffer solution. Why are they included?

PRACTICAL ACTIVITY 3

4 What is meant by 'target DNA'?

5 Why has DNA polymerase been included in the buffer solution?

6 The DNA polymerase is not human polymerase, but rather *Taq* polymerase derived from the bacterium *Thermus aquaticus*, which inhabits hot springs. Outline the properties of *Taq* polymerase that make it a useful tool in PCR.

7 Describe what happens at the following temperatures.

a 95°C

b 50°C

c 72°C

8 How many copies of target DNA fragments are made after:

a three PCR cycles

b four PCR cycles

c five PCR cycles

d 30 PCR cycles

9 Describe the outcome of PCR.

 ISBN 978 0 6557 0026 5

CONCLUSIONS

10 Summarise the process of PCR using a maximum of two sentences.

11 Suggest how the process of PCR is important to gene technologies such as gel electrophoresis and DNA profiling, which analyse DNA recovered at crime scenes and are used to investigate relationships between individuals.

PRACTICAL ACTIVITY 4

Case study • Literature review

Genome editing using CRISPR-Cas9

Suggested duration: Part A—60 minutes; Part B—120 minutes

INTRODUCTION

CRISPR is an abbreviation for 'clustered regularly interspaced short palindromic repeats'. That means just what it says—short DNA sequences that are palindromic (read the same both forwards and backwards) and appear in clusters with regular spacing between them. The spacers contain unique DNA segments—they are not repeats, but different from each other.

Why has CRISPR been the focus of so much attention in recent times if it is just DNA? And why is it so controversial?

CRISPR is a special kind of DNA found in bacteria, which use it to combat viral infections. In association with a protein called Cas9, scientists have been able to construct a bacterial CRISPR-Cas9 system that can edit DNA sequences in living cells. The possibilities for the application of this technology are profound. The genomes of organisms could be manipulated to confer desirable characteristics in plants and animals, for example, to improve crop yield and quality in agriculture; to combat disease by altering bacteria and viruses; and to bypass genetically inherited disorders by correcting faulty genes.

In this activity you will find out more about CRISPR-Cas9 and explore some possible applications of the technology. You will consider the issues raised by the possible applications of CRISPR-Cas9 and make predictions about potential long-term consequences of some of its applications.

AIM

- To investigate the genome editing tool CRISPR-Cas9.
- To consider some issues related to the applications of genome editing.

BACKGROUND

Figure 3.1.32 shows a simplified representation of the CRISPR-Cas9 molecule. Note the following features:

- the Cas9 enzyme
- a sequence of RNA called CRISPR RNA (crRNA)
- a sequence of RNA called trans-activating CRISPR RNA (tracrRNA)
- a sequence of RNA called guide RNA (gRNA).

The diagram on the right in Figure 3.1.32 shows the crRNA joined to the tracrRNA. This combined RNA complex is referred to as the guide RNA (gRNA).

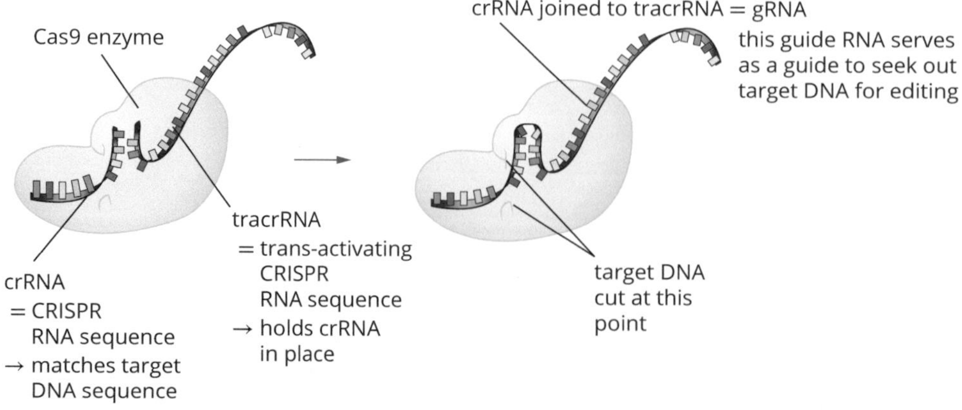

Figure 3.1.32 CRISPR-Cas9 molecule (left) and guide RNA (gRNA) (right)

 ISBN 978 0 6557 0026 5

PRACTICAL ACTIVITY 4

When a CRISPR-Cas9 system has been created to match a target sequence of DNA, it is ready to be introduced into affected cells for DNA editing. Figure 3.1.33 shows the CRISPR-Cas9 system working through the steps of editing a sequence of target DNA. DNA editing using the CRISPR-Cas9 system could include correcting a faulty gene by replacing a nucleotide, adding a DNA sequence or deleting a DNA sequence.

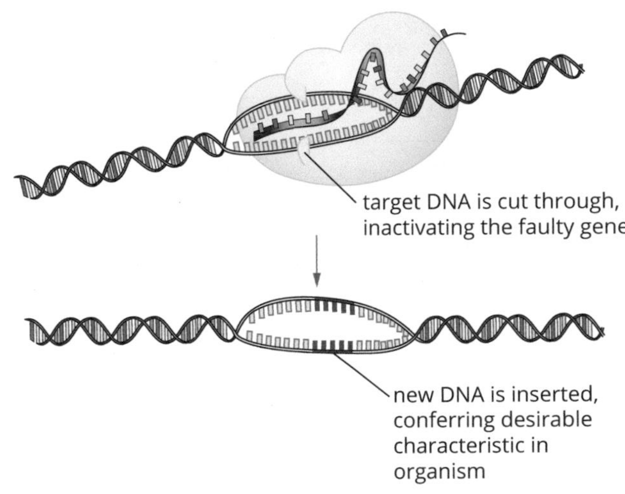

Figure 3.1.33 The CRISPR-Cas9 system editing a sequence of target DNA

PART A • CRISPR-Cas9 AND ITS APPLICATIONS

Use the internet to locate and watch a short animation outlining the CRISPR-Cas9 system. Watch the video a couple of times, stopping and starting as needed. Useful keywords/phrases for your search might include CRISPR-Cas9 system; What is CRISPR?; explaining CRISPR.

1 Explain what is meant by each of the following terms.

a target DNA

____

b crRNA

____

c tracrRNA

____

d gRNA

____

2 In the CRISPR-Cas9 complex, describe the role of:

a the Cas9 enzyme

____

b guide RNA

____

3 **a** CRISPR-Cas9 technology has many potential applications. For example, it could be used to edit mosquito DNA as a means of controlling mosquito populations in malaria zones, with a view to eradicating malaria. Suggest other possible applications of CRISPR-Cas9.

____

____

____

____

b Share your response with other students, adding to your list of applications where relevant.

ISBN 978 0 6557 0026 5

4 Outline how applying CRISPR-Cas9 technology to edit somatic cells is different from applying it to edit germ cells.

5 Describe potential benefits from the development and application of CRISPR-Cas9 as a genome-editing tool.

PART B • ISSUES AROUND CRISPR-Cas9

The possibilities of CRISPR-Cas9 technology come hand-in-hand with some far-reaching implications and issues that are the subject of a great deal of controversy and ethical debate. Such issues fall under various contexts including social, cultural, economic, political and legal.

6 Suggest issues raised in the event that CRISPR-Cas9 technology were to be used as a tool in mosquito control to eradicate malaria.

7 Consider the potential applications of CRISPR-Cas9 technology recorded in question 3a.

  a Select one of these issues for further consideration. Identify the issue.

  b Use resources to gather 'big picture' information about the interplay between this potential CRISPR-Cas9 application in society in terms of:

  - a social context
  - a cultural context
  - an economic context.

  Include information about the following key points:

  - Who are the stakeholders in the community?
  - What are the perspectives held by the different stakeholders in terms of the technology and its outcomes?
  - What arguments would they be likely to put forward to support their positions?
  - How does the context influence the applications of the technology?
  - What are the advantages and disadvantages of the technology in this application?
  - What are the potential long-term outcomes?

 ISBN 978 0 6557 0026 5

PRACTICAL ACTIVITY 4

Prepare a mind map summarising the information you have gathered.

8 A newspaper editorial is an article in which the editor takes in the information around an important issue affecting our society and outlines their opinion. Use all you have learned about your selected CRISPR-Cas9 application to write your own newspaper editorial on the application.

CONCLUSIONS

9 Use one or two sentences to summarise how the CRISPR-Cas9 system works to edit genes.

10 Write a short response to the information you have learned about CRISPR-Cas9 technology and its applications.

EXAM QUESTIONS

Multiple-choice questions

Question 1 VCE Biology 2016 (A) 7

In animal cells, tight junctions are multi-protein complexes that mediate cell-to-cell adhesion and regulate transport through the extracellular matrix. Proteins that form these complexes are made within the cell.

One pathway for the production of protein for these junctions is

A. nucleus—ribosome—Golgi apparatus—vesicle—endoplasmic reticulum.

B. nucleus—ribosome—endoplasmic reticulum—vesicle—Golgi apparatus.

C. nucleus—vesicle—endoplasmic reticulum—Golgi apparatus—ribosome.

D. nucleus—vesicle—Golgi apparatus—ribosome—endoplasmic reticulum.

Use the following information to answer questions 2 and 3. VCE Biology 2015 (A) 3 and 4

The diagram below represents part of a DNA molecule.

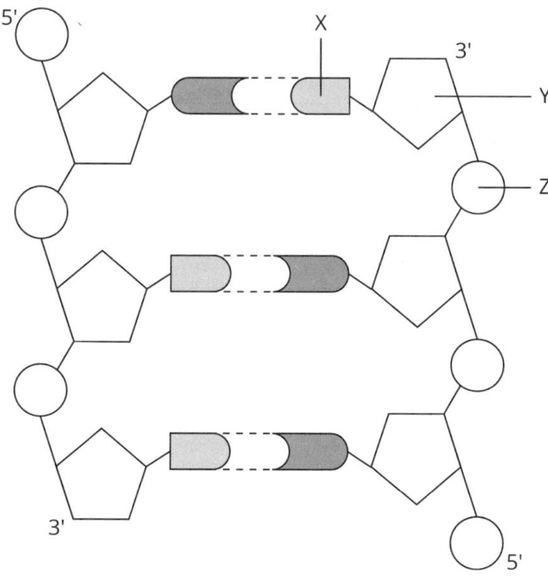

Question 2 VCE Biology 2015 (A) 3

A single DNA nucleotide is shown by sub-unit(s)

A. X alone.

B. X and Y together.

C. Y and Z together.

D. X, Y and Z together.

Question 3 VCE Biology 2015 (A) 4

A feature of DNA that can be seen in the diagram above is

A. the anti-parallel arrangement of the two strands of nucleotides.

B. the process of semi-conservative replication.

C. its ribose sugar–phosphate backbone.

D. its double-helix structure.

ISBN 978 0 6557 0026 5

EXAM QUESTIONS

Question 4 VCE Biology 2017 (A) 1

Consider the structure and functional importance of proteins.

Which one of the following statements about proteins is correct?

A. A change in the tertiary structure of a protein may result in the protein becoming biologically inactive.

B. Proteins with a quaternary structure will be more active than proteins without a quaternary structure.

C. Two different proteins with the same number of amino acids will have identical functions.

D. Denaturation will alter the primary structure of a protein.

Question 5 VCE Biology 2017 (A) 1

The *lac* operon was originally identified in *Escherichia coli*. The *lac* operon has three structural genes: *lacZ*, *lacY* and *lacA*. The *lacZ* gene codes for the production of the enzyme β-galactosidase, which catalyses the breakdown of lactose into glucose and galactose. Below is a diagram that shows the order of the genes found in the *lac* operon. The dots represent the DNA nucleotides between the genes.

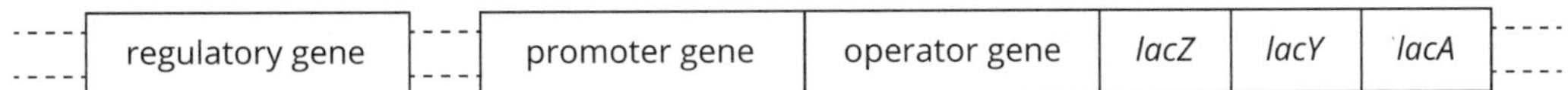

To begin transcription of the three structural genes, RNA polymerase needs to bind to the

A. operator gene.

B. promoter gene.

C. regulatory gene.

D. structural genes.

Question 6 VCE Biology 2018 (A) 3

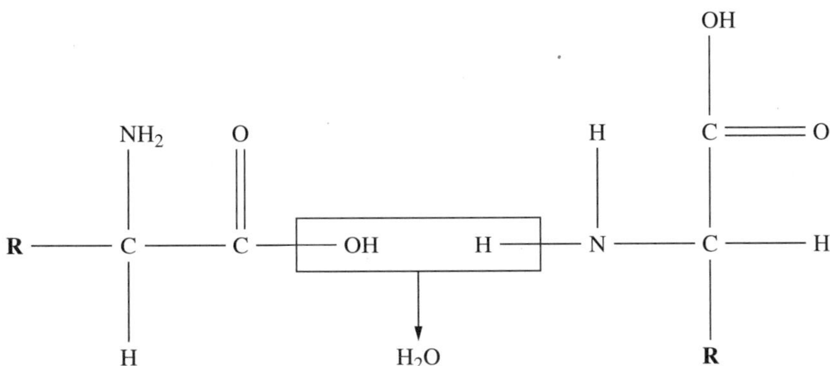

The diagram above represents adjacent amino acids being joined together.

The joining of adjacent amino acids

A. results in the formation of a nucleic acid.

B. is an energy-releasing reaction.

C. is catalysed by DNA ligase.

D. is a condensation reaction.

EXAM QUESTIONS

Question 7 VCE Biology 2014 (A) 24

Casein is a major protein found in mammalian milk.

When the mammals are producing milk, the pathway for the production of casein can be represented as shown in the diagram below on the left.

When the mammals are not producing milk, the pathway can be represented as shown in the diagram below on the right.

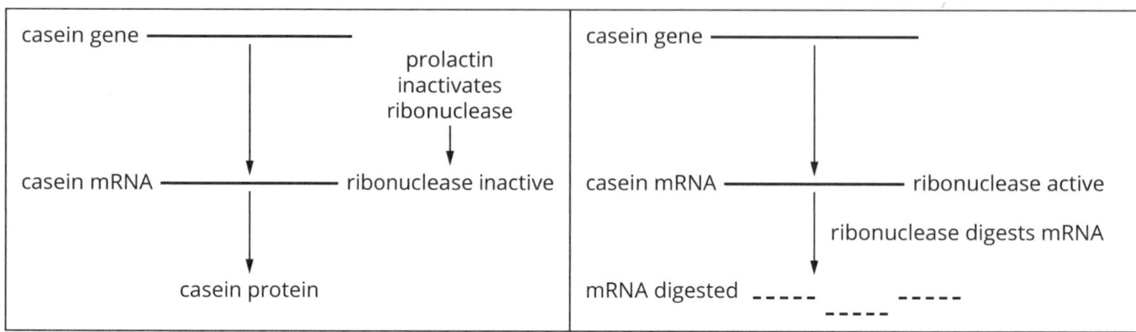

Which one of the following conclusions can be made from the information above?

A. Ribonuclease has the effect of turning on the casein gene.

B. Casein is a repressor protein for milk production in mammals.

C. The hormone prolactin allows for the expression of the casein gene.

D. Mammals produce milk only in the absence of the hormone prolactin.

Question 8 VCE Biology 2018 (A) 28

The diagram below represents a method of DNA manipulation.

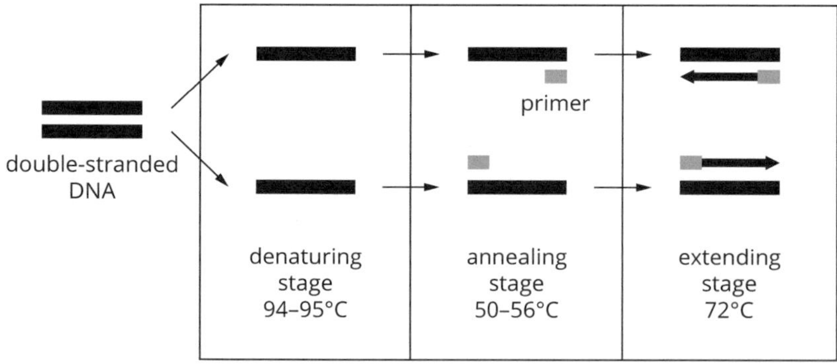

The method represented is

A. gel electrophoresis.

B. DNA transformation.

C. bacterial transformation.

D. polymerase chain reaction.

 ISBN 978 0 6557 0026 5

EXAM QUESTIONS

Question 9 VCE Biology 2018 (A) 34

Genetic testing can be used to test for the allele for Huntington's disease (HD). The onset of HD predominantly occurs in adulthood.

Eight individual family members were tested for the HD allele. The diagram below shows the electrophoresis gel results of a test for the presence of the allele. Individuals 4 and 8 have been diagnosed with the disease.

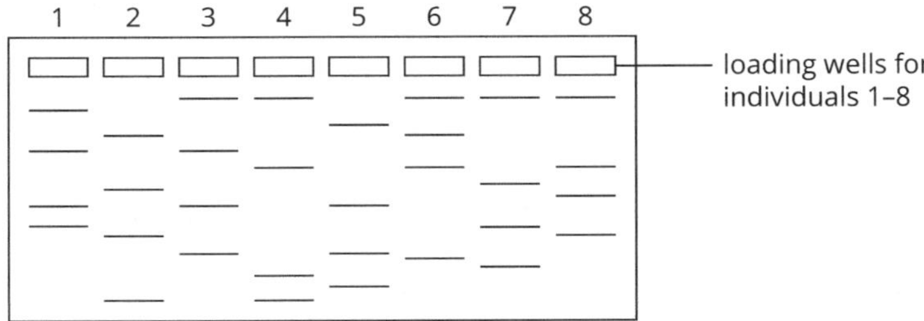

Which other individual is likely to suffer from HD now or in the future?

A. 1

B. 2

C. 5

D. 6

Question 10 VCE Biology 2017 (A) 34

The process known as polymerase chain reaction (PCR) involves repeated cycles made up of several steps.

During PCR the

A. first step in each cycle is to anneal primers to the DNA at a low temperature.

B. temperature must be lowered to 37°C before the beginning of each cycle.

C. second step in each cycle is to heat the DNA to a high temperature.

D. final step of each cycle involves the use of DNA polymerase.

Short-answer questions

Question 1 (6 marks) VCE Biology 2008 (1) (B) 2

There are structural differences between molecules of DNA and RNA.

a. Outline two of these differences by completing the following table. 2 marks

	DNA	RNA
Difference 1		
Difference 2		

b. Name one kind of RNA and state its function. 1 mark

Type of RNA: ______________________________

Function: ______________________________

____

EXAM QUESTIONS

Proteins may be classified as fibrous or globular depending on their three-dimensional shape.

In fibrous proteins, the polypeptide chains are arranged in parallel to form long fibres or sheets. In globular proteins, the polypeptide chains are folded into compact spherical or globular shapes.

c. Name the subunit of a polypeptide. 1 mark

Keratin, found in fingernails and claws, is an example of a fibrous protein.

d. Name another example of a fibrous protein and briefly outline its function. 1 mark

e. Describe a distinctive property of a fibrous protein and explain how this property is due to the arrangement of its polypeptides. 1 mark

Question 2 (6 marks) VCE Biology 2012 (1) (B) 3

Human insulin is a macromolecule composed of two amino acid chains. The chains are connected by disulfide bonds.

a. To what group of macromolecules does insulin belong? 1 mark

Insulin found in other animals varies from human insulin.

The following table compares all the differences seen in the primary structure of human, cow, pig and sheep insulin.

	Amino acid position number within	
	Alpha chain	**Beta chain**
	-8 - 9 - 10-	**-30-**
human	-thr - ser - ile-	thr
cow	-ala - ser - val-	ala
pig	-thr - ser - ile-	ala
sheep	-ala - gly - val-	ala

b. What is meant by the term 'primary structure' of the insulin macromolecule? 1 mark

Humans with diabetes take insulin injections to maintain their health.

c. If supplies of human insulin were not available, which one of the other three animals listed in the table would be the best source of insulin? Explain your reason for choosing this particular animal. 2 marks

Animal:

Explanation:

 ISBN 978 0 6557 0026 5

EXAM QUESTIONS

The table below contains the genetic code for protein production.

first letter	second letter: A	second letter: G	second letter: T	second letter: C	third letter
A	AAA, AAG: phe AAT, AAC: leu	AGA, AGG, AGT, AGC: ser	ATA, ATG: tyr ATT: Stop ATC: Stop	ACA, ACG: cys ACT: Stop ACC: trp	A G T C
G	GAA, GAG, GAT, GAC: leu	GGA, GGG, GGT, GGC: pro	GTA, GTG: his GTT, GTC: gln	GCA, GCG, GCT, GCC: arg	A G T C
T	TAA, TAG, TAT: ile TAC: met Start	TGA, TGG, TGT, TGC: thr	TTA, TTG: asn TTT, TTC: lys	TCA, TCG: ser TCT, TCC: arg	A G T C
C	CAA, CAG, CAT, CAC: val	CGA, CGG, CGT, CGC: ala	CTA, CTG: asp CTT, CTC: glu	CCA, CCG, CCT, CCC: gly	A G T C

d. Use the information in the table **to explain**

i. the different sequence of nucleotides in humans and cows with respect to the DNA coding for the amino acid at position 30. 1 mark

ii. whether the sequence of nucleotides in DNA coding for the amino acid at position 30 will be identical in cows, pigs and sheep. 1 mark

Question 3 (7 marks) VCE Biology 2017 (B) 9

Scientists use recombinant bacterial plasmids as vectors to transform bacteria for a range of purposes in research and biotechnology.

a. What is meant by the term 'vector' in the context given? 1 mark

EXAM QUESTIONS

A particular bacterial plasmid contains recognition sites for the restriction enzymes EcoRI, HindIII and BamHI, along with two antibiotic-resistant genes, ampicillin resistance (*amp*) and tetracycline resistance (*tcl*), and an origin of replication (ORI).

The diagram below shows the positions of these recognition sites and antibiotic-resistant genes as well as the position of the origin of replication within this plasmid.

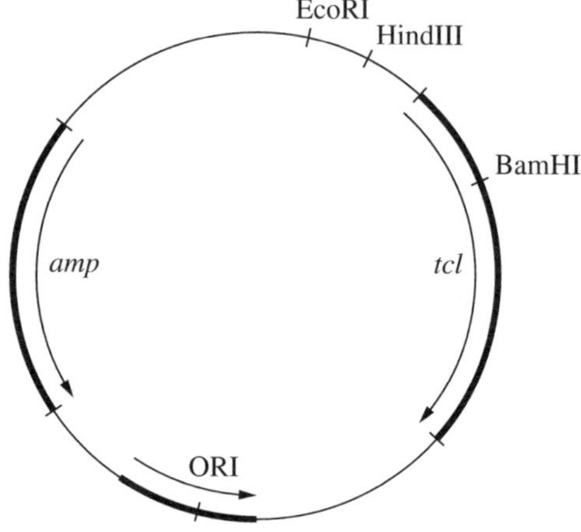

One purpose of using recombinant bacterial plasmids is to produce bacteria capable of synthesising human protein.

b. The restriction enzyme BamHI was used to help insert a gene coding for a human protein into this plasmid.

i. Describe how restriction enzymes such as BamHI are used to help insert a gene coding for a human protein into this plasmid. 2 marks

ii. Draw and label a diagram in the space below to show the position of the human gene in this plasmid when BamHI is used. Include the position of the recognition sites for the restriction enzymes EcoRI, HindIII and BamHI on the plasmid. 1 mark

 ISBN 978 0 6557 0026 5

c. After the scientists had carried out the steps required to make plasmids with the inserted human gene, these plasmids were mixed with a culture of bacteria. This mixture was treated so that these plasmids would move into the bacterial cells. Not all bacteria took up these plasmids.

Explain how scientists use antibiotics to identify which of the bacterial cells have been successfully transformed with plasmids carrying the human gene. 3 marks

Question 4 (8 marks) VCE Biology 2019 (B) 8

The Genomics Health Futures Mission will run a $32 million trial, starting in 2019, to screen over 10 000 couples who are in early pregnancy or who are planning to have a baby. Using a blood test, individuals will be screened for 500 severe or deadly recessive gene mutations.

Couples will be told they have a genetic mutation if both individuals in the couple carry the same mutation. The trial may lead to a population-wide carrier screening program. The researchers will evaluate cost effectiveness, psychological impact, ethics and barriers to screening. It is anticipated that future tests will be free of charge.

a. Give an example of a disorder or disease that can be detected by genetic screening. 1 mark

b. The blood sample from an individual will provide researchers with only a small amount of DNA. Polymerase chain reaction (PCR) will be used to amplify the DNA.

Describe what happens in each of the three stages of PCR. The stages must be given in the order in which they occur. 3 marks

Stage	What is happening at this stage
1	
2	
3	

c. Once the DNA has been amplified, it can be loaded into a well of an agarose gel.

Discuss **three** factors that affect the migration of DNA fragments through the agarose gel during gel electrophoresis. 3 marks

d. The test may find that a couple who were planning to have a baby or who were already pregnant both carry the same severe or deadly mutation.

Describe one ethical and one social issue/implication that could arise from this finding. 3 marks

	Issue/Implication
Ethical	
Social	

Question 5 (8 marks) VCE Biology 2018 (B) 10

Should we grow GM crops?

by Mary Nguyen

More than 25 years after genetically modified (GM) food first appeared, growing GM crops remains a hotly debated topic. Some people argue that GM crops are the only way to feed the growing world population and to minimise environmental harm. Other people express different views.

Bt cotton is a type of cotton that contains two genes from a soil bacterium, *Bacillus thuringiensis*, enabling it to produce insect-resistant proteins. Australian farmers of Bt cotton use only 15% of the quantity of the insecticide that was once needed to protect their cotton crops*. However, Bt cotton is not as resistant to the main insect pest of cotton crops, *Helicoverpa*, as it has been in the past*.

In Australia, Bt cotton is picked by machine, but in India, it is picked by hand. Workers in India have developed skin allergies, which have been attributed to Bt cotton proteins.

Traditionally, farmers have saved money by keeping seed from one year's crop to plant the following year. However, it is illegal for farmers to keep Bt cotton seeds because these seeds have been declared the legal property of the company Monsanto. Every year, cotton farmers must buy more seeds from Monsanto.

Unlike Monsanto, the company that produces the GM food crop Golden Rice allows farmers to replant the rice they harvested the previous year. By inserting a gene from the bacteria *Erwinia uredovora* and another from a daffodil, *Narcissus pseudonarcissus*, into white rice, scientists produced Golden Rice—a rice variety containing higher levels of vitamin A†. People who eat Golden Rice avoid vitamin A deficiency. Trials conducted in several countries have shown that Golden Rice is safe to eat‡.

References: *CSIRO, 'Cotton pest management', case study, <www.csiro.au>; †JA Paine et al., 'Improving the nutritional value of Golden Rice through increased pro-vitamin A content', *Nature Biotechnology*, 23, 27 March 2005, pp. 482–487; ‡A Coghlan, 'Golden Rice gets approval in the US', *New Scientist*, magazine issue 3180, 2 June 2018

 ISBN 978 0 6557 0026 5

EXAM QUESTIONS

a. Bt cotton and Golden Rice are genetically modified organisms but are they also transgenic organisms? Support your response with evidence from the article above. 3 marks

b. How can planting a Bt cotton crop lead to an increase in crop yield? 1 mark

c. Using information from the article, complete the table below by describing one social and one biological implication relevant to the use of Bt cotton and Golden Rice. The same implication should **not** be used twice. 4 marks

	Social implication	Biological implication
Bt cotton		
Golden Rice		

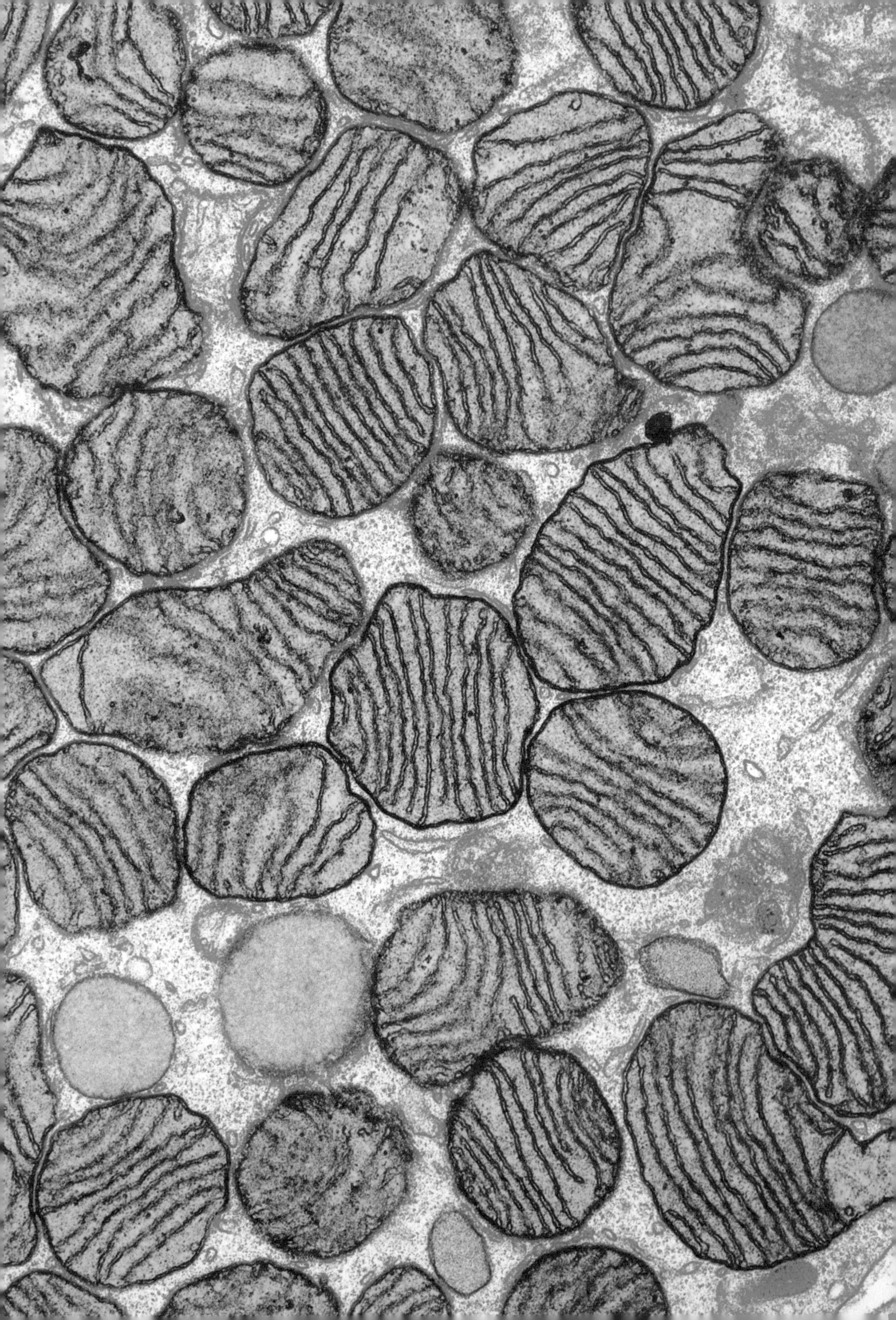

UNIT 3

How do cells maintain life?

AREA OF STUDY 2

How are biochemical pathways regulated?

Outcome 2

On completion of this unit the student should be able to analyse the structure and regulation of biochemical pathways in photosynthesis and cellular respiration, and evaluate how biotechnology can be used to solve problems related to the regulation of biochemical pathways.

Key knowledge

Regulation of biochemical pathways in photosynthesis and cellular respiration

- the general structure of the biochemical pathways in photosynthesis and cellular respiration from initial reactant to final product
- the general role of enzymes and coenzymes in facilitating steps in photosynthesis and cellular respiration
- the general factors that impact on enzyme function in relation to photosynthesis and cellular respiration: changes in temperature, pH, concentration, competitive and non-competitive enzyme inhibitors

Photosynthesis as an example of biochemical pathways

- inputs, outputs and locations of the light dependent and light independent stages of photosynthesis in C_3 plants (details of biochemical pathway mechanisms are not required)
- the role of Rubisco in photosynthesis, including adaptations of C_3, C_4 and CAM plants to maximise the efficiency of photosynthesis
- the factors that affect the rate of photosynthesis: light availability, water availability, temperature and carbon dioxide concentration

Cellular respiration as an example of biochemical pathways

- the main inputs, outputs and locations of glycolysis, Krebs cycle and electron transport chain including ATP yield (details of biochemical pathway mechanisms are not required)
- the location, inputs and the difference in outputs of anaerobic fermentation in animals and yeasts
- the factors that affect the rate of cellular respiration: temperature, glucose availability and oxygen concentration

Biotechnological applications of biochemical pathways

- potential uses and applications of CRISPR-Cas9 technologies to improve photosynthetic efficiencies and crop yields
- uses and applications of anaerobic fermentation of biomass for biofuel production.

Regulation of biochemical pathways

The **metabolism** of an organism is the sum of all the chemical reactions that occur within its cells. This includes the energy-transforming reactions of cells such as the production of organic molecules, and the breakdown, recycling and excretory processes. These biochemical processes are universal; that is, they occur in the cells of all living organisms to ensure the survival of the individual.

Enzymes are biological catalysts—they increase the rate of biochemical reactions in cells, for example, the chemical reactions involved in cellular respiration and photosynthesis.

ENZYME FEATURES

Enzymes are usually denoted by the suffix 'ase', for example, maltase, lactase, protease, amylase and lipase. Enzymes:

- are composed of protein
- are **substrate** specific—they **catalyse** a chemical reaction involving a particular substrate molecule, and not any other. There are two theories used to explain how enzymes interact with their substrates: the **lock-and-key model** and the **induced-fit model** (Figures 3.2.1a and 3.2.1b)
- take part in chemical reactions but are not used up or changed by the process. They are released at the end of a reaction and so are available to be used over and over again
- have optimal conditions under which they work most efficiently; they will catalyse a reaction so that maximum product is produced per unit of time. For example, the optimal conditions for digestive enzymes in the duodenum of humans is 37°C and pH 8
- are sensitive to factors such as temperature and pH. When these conditions are not optimal, the activity of enzymes is reduced. Extremes of these factors may lead to enzymes becoming **denatured**. When this happens the enzyme cannot recover its function because the shape of its **active site** has been permanently altered (Figure 3.2.2).

Lock-and-key model
substrate molecules
product
active site
enzyme
enzyme–substrate complex

Figure 3.2.1a Model of enzyme 'lock-and-key' operation

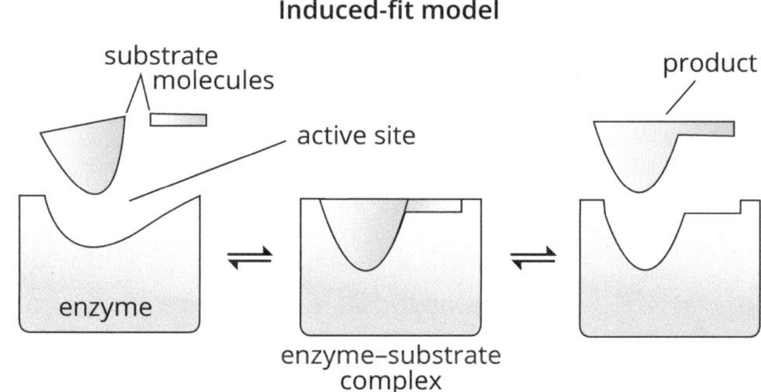

Figure 3.2.1b Model of enzyme 'induced-fit' operation

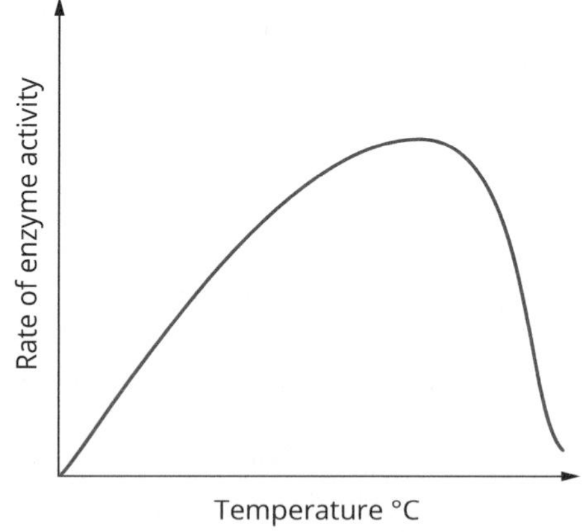

Figure 3.2.2 Rate of enzyme activity continues to increase with increasing temperature. Excessive temperature denatures enzymes and activity ceases.

 ISBN 978 0 6557 0026 5

In summary, enzymes regulate biochemical pathways, acting on substrate molecules (**reactants**) to form a final product. During this process, enzymes interact with substrate molecules in a series of intermediate steps that involve the formation of enzyme–substrate complexes. It is in these intermediate steps that chemical bonds are either set to create larger, more complex molecules, or broken down resulting in smaller, simpler molecules.

The rate of enzyme activity is also dependent upon the:

- concentration of substrate—the higher the concentration of the substrate, the greater the rate of interaction between substrate molecules and enzymes, leading to an increased rate of reaction
- concentration of enzyme—the more enzyme available to catalyse a reaction, the more rapidly the reaction will proceed until all enzyme molecules are fully engaged in the reaction (Figure 3.2.3).

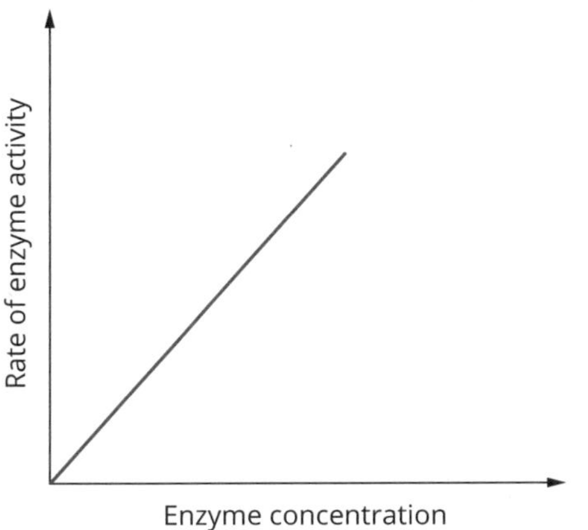

Figure 3.2.3 Rate of enzyme activity increases with increasing enzyme concentration

INHIBITION OF ENZYME ACTIVITY

The action of enzymes is also influenced by the presence of chemical competitors that interact with an enzyme by binding to the active site or some other location on the enzyme molecule. Such molecules are called enzyme inhibitors because they decrease an enzyme's ability to interact with the intended substrate, thereby decreasing or halting the chemical pathway. Enzyme inhibitors are described as competitive or non-competitive.

Competitive inhibition

Competitive inhibition occurs when the shape of the inhibitor is similar to the shape of the substrate that normally binds to the active site of an enzyme. Due to their similar shapes, such inhibitors are able to bind to the active site of the enzyme, blocking the substrate from binding to the site and halting the reaction. The antibiotic penicillin is an example of a competitive enzyme inhibitor that is also irreversible. Strong covalent bonds mean the competitor is permanently attached to the active site of an enzyme involved in construction of the bacterial wall. This outcome renders pathogenic bacteria unable to survive, hence the use of penicillin in medicine.

Some competitive enzyme inhibitors form weak, non-permanent bonds with the active site of enzymes. Reversible enzyme inhibition has an important regulatory role in cells. Inhibitors are in place when enzymes are not required and dissociate from the enzyme when the enzyme product is needed.

Non-competitive inhibition

Non-competitive inhibition occurs when an inhibitor binds to an enzyme at a different location from the active site. Non-competitive inhibitors are so named because the inhibitor does not fit or compete with substrate molecules for the enzyme's active site. Non-competitive inhibitors change the shape of the enzyme molecule, altering the active site so that substrate molecules are no longer a neat fit. As a result, enzyme activity is no longer optimal.

Activation energy is the energy expended to initiate a reaction; even **catabolic reactions** require an initial input of energy to start the reaction. Enzymes lower the activation energy, making it easier for the reaction to proceed.

Coenzymes are small organic molecules that are important to the normal functioning of enzymes. **Adenosine triphosphate (ATP)** is a coenzyme that plays a critical role in the transfer of energy in reactions such as cellular respiration. NADH and NADPH represent coenzymes involved in the cycling of protons and electrons in energy transformations including cellular respiration and photosynthesis. Acetyl coenzyme A is another example of a coenzyme important in cellular respiration.

Cofactors are inorganic ions important to the normal functioning of enzymes, for example, Mg^{++}.

Biochemical pathways: photosynthesis and cellular respiration

Living organisms require energy for growth, movement, repair of damaged tissue and reproduction. Organisms that produce their own organic compounds are called **autotrophs**. Autotrophs can be photosynthetic (plants) or chemosynthetic (some prokaryotes). Organisms that obtain their organic compounds from other organisms are called **heterotrophs**. Examples of heterotrophs are fungi and animals. Different kinds of energy transformations are involved.

PHOTOSYNTHESIS

The ultimate source of energy for living things is the Sun. **Photosynthesis** is the process in which green plants use **chlorophyll** to trap light energy and use it to combine water (H_2O) and carbon dioxide (CO_2) to produce energy-rich organic compounds (**glucose**). Oxygen is a by-product (Figure 3.2.4).

Photosynthesis involves the synthesis of biomacromolecules and is therefore described as an **anabolic reaction**. Since this process requires an input of energy, it is an **endergonic reaction**.

A word equation for photosynthesis is:

$$\text{carbon dioxide + water} \xrightarrow[\text{chlorophyll}]{\text{light}} \text{glucose + oxygen}$$

A balanced chemical equation for photosynthesis is:

$$6CO_2 + 12H_2O \xrightarrow[\text{chlorophyll}]{\text{light}} C_6H_{12}O_6 + 6O_2 + 6H_2O$$

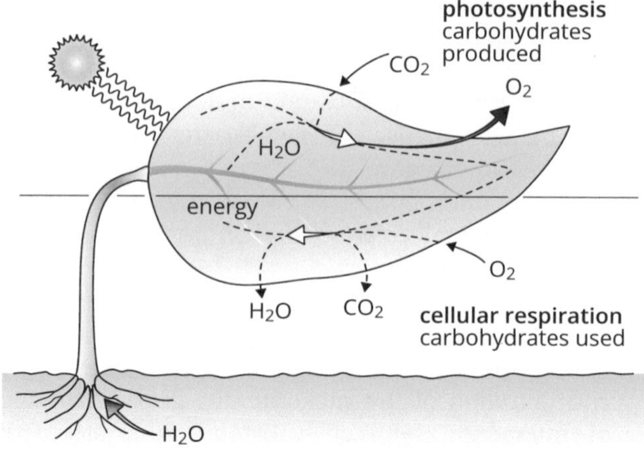

Figure 3.2.4 Summary of photosynthesis and cellular respiration

Photosynthesis takes place in the chloroplasts and occurs in steps called the **light-dependent reaction** and the **light-independent reaction**. The light-dependent reaction occurs in the presence of light and the light-independent reaction occurs in the absence of light. These reactions take place at different sites in the chloroplasts: the light-dependent reaction occurs in **grana** and the light-independent reaction occurs in the **stroma** (Figure 3.2.5 and Table 3.2.1).

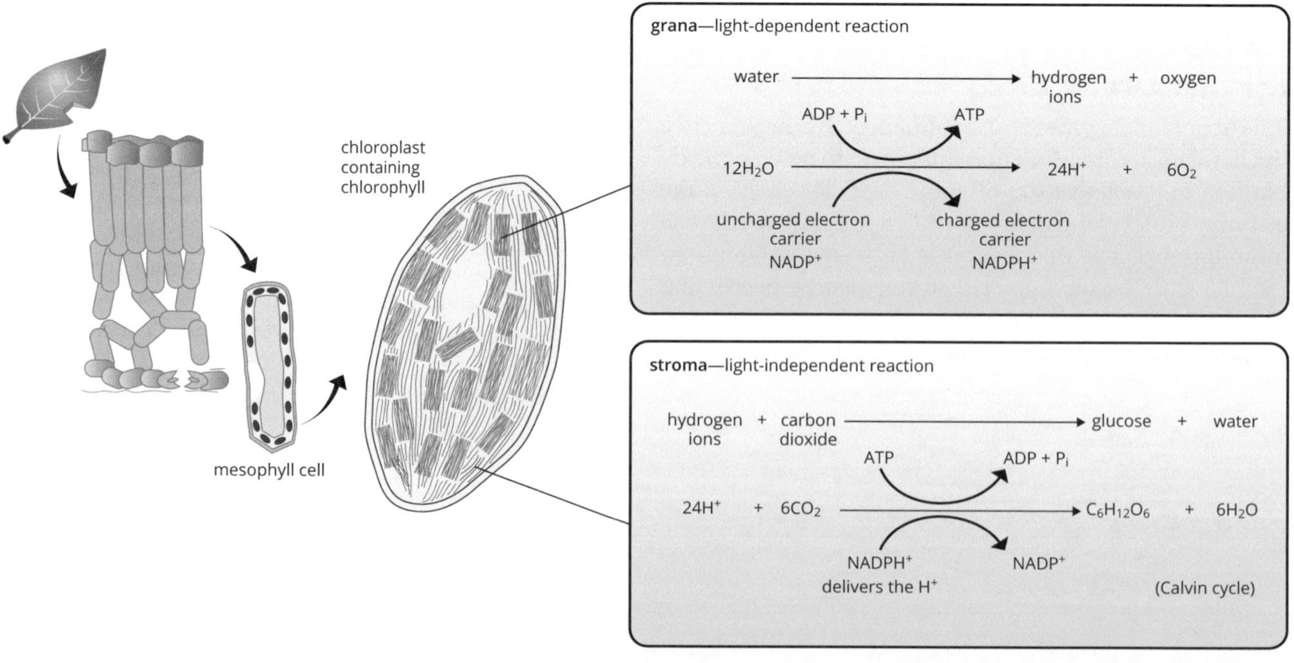

Figure 3.2.5 The stages of photosynthesis take place at different sites in the chloroplasts—the light-dependent reaction occurs in the grana and the light-independent reaction occurs in the stroma.

ISBN 978 0 6557 0026 5

KEY KNOWLEDGE

The light-independent reaction in **C_3 plants** is also called the **Calvin cycle**. C_3 refers to the three-carbon carbohydrate molecules that are produced in this process, precursors to the six-carbon glucose molecule that is the final product.

Table 3.2.1 Photosynthesis—inputs and outputs

	Inputs	Outputs
Light-dependent stage	• water • $NADP^+$ • ADP + P_i	• O_2 • NADPH • ATP
Light-independent stage	• CO_2 • NADPH • ATP	• $C_6H_{12}O_6$ (glucose) • $NADP^+$ • water • ADP + P_i

The light compensation point, shown in Figure 3.2.6, is the light intensity at which the rate of carbon dioxide produced by cellular respiration is equal to the rate of carbon dioxide uptake in photosynthesis. This occurs when the rates of cellular respiration and photosynthesis are the same.

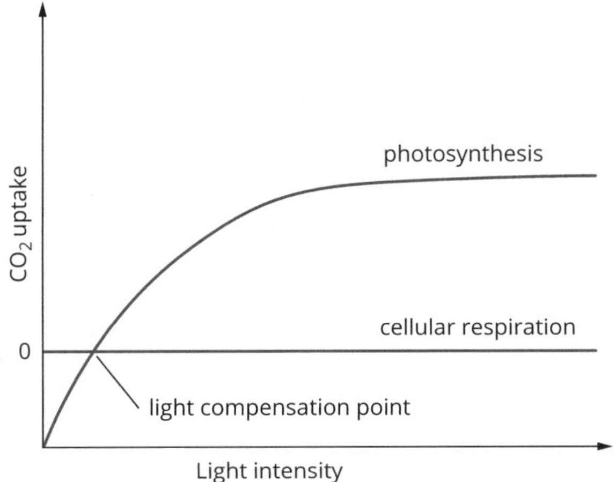

Figure 3.2.6 Light compensation point

The factors affecting the rate of photosynthesis are summarised in Figure 3.2.7 and Table 3.2.2.

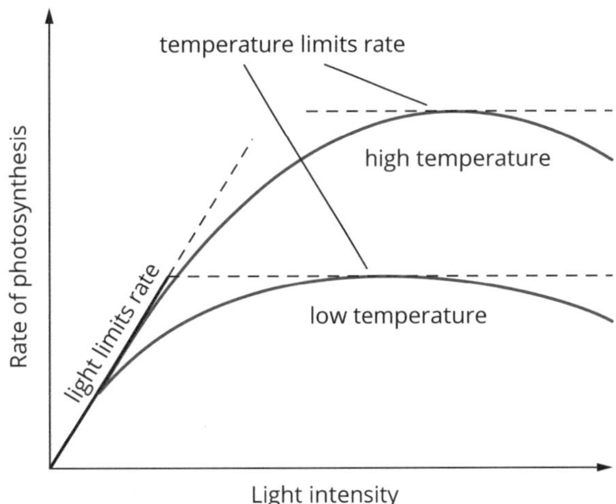

Figure 3.2.7 Effect of temperature and light intensity on rate of photosynthesis

Table 3.2.2 Factors that affect the rate of photosynthesis in green plants

Factor	Effect
carbon dioxide concentration	The higher the concentration gradient between the intercellular spaces and the external environment, the greater the diffusion rate of carbon dioxide into leaves via stomata, and, therefore, the greater the rate of photosynthesis.
light intensity	The greater the light intensity, the greater the rate of photosynthesis, until the point at which other factors limit the process, e.g. high temperatures result in closed stomata.
temperature	Increasing temperature results in an increased rate of photosynthesis, until the optimal temperature for photosynthetic enzymes is reached; further increases in temperature result in a decrease in the photosynthetic rate.
water	Water is a reactant in photosynthesis; it is also important in the turgidity of guard cells, keeping stomata open for the entry of carbon dioxide. Reduced water availability decreases the rate of photosynthesis.
chlorophyll	Light-trapping pigment; larger amounts of chlorophyll mean more sunlight is harnessed, thereby increasing the rate of photosynthesis.
oxygen	Not directly involved in photosynthesis; however, high oxygen concentrations reduce the rate of carbon fixation in photosynthesis.

KEY KNOWLEDGE

Photosynthesis in C_3, C_4 and CAM plants

Rubisco (short for ribulose-1,5-biphosphate carboxylase/oxygenase) is an enzyme that plays a critical role in photosynthesis. Found in the stroma of chloroplasts in C_3 plants, Rubisco is involved in processing atmospheric carbon dioxide into carbohydrates in a process called **carbon fixation**. Specifically, the enzyme is involved in the Calvin cycle where it catalyses the reaction in which a carbon dioxide molecule is added to a five-carbon sugar called ribulose phosphate, forming a six-carbon molecule that is split into two three-carbon molecules of phosphoglyceric acid (PGA). The PGA molecules are eventually combined to produce energy-rich glucose.

Some species of plants adapted to hot environments have different photosynthetic pathways. These pathways are solutions for plants that must have their stomata closed for much of the day to avoid excessive water loss and so have less time to acquire carbon dioxide for photosynthesis. **C_4 plants**, such as sugar cane and sorghum, are more efficient at fixing carbon than C_3 plants, capturing more carbon dioxide per unit time. The first stable carbon compound produced contains four carbon atoms, hence the term C_4 photosynthesis.

CAM plants, such as pineapple, only have their stomata open during the night. The carbon dioxide they collect is converted to crassulacean acid in a process called **crassulacean acid metabolism** (or CAM photosynthesis). During the day, when the stomata are closed, the carbon dioxide is released from storage and is available to take part in normal C_3 photosynthesis.

CELLULAR RESPIRATION

In contrast to photosynthesis, **cellular respiration** involves the chemical breakdown of glucose in cells for the release of energy. This involves the uptake of O_2 and the release of CO_2.

Some autotrophic bacteria use inorganic materials to build organic compounds without the involvement of sunlight. This is called chemosynthesis.

ATP is the immediate source of energy for cells. It is produced in a series of chemical reactions that involves the breakdown of organic molecules. The useable energy of ATP is contained in the phosphate bonds of the molecule. The cycling of ATP and **adenosine diphosphate (ADP)** means that energy continues to be available for use in the cell (Figure 3.2.8).

Cells access the energy available in organic molecules through **glycolysis** (anaerobic), and either cellular respiration (aerobic) or **anaerobic fermentation**.

Cellular respiration is the process in which complex organic compounds are broken down to release energy (ATP). Water and carbon dioxide are by-products.

Cellular respiration consists of three interconnected biochemical pathways:

- glycolysis
- the **Krebs cycle** (also called the citric acid cycle)
- the **electron transport chain**.

Aerobic cellular respiration begins with glycolysis, which occurs in the cytoplasm of the cell. It is then followed by the Krebs cycle and the electron transport chain, both of which occur in the **mitochondria** (Figure 3.2.9).

Table 3.2.3 summarises the inputs and outputs during the three stages of cellular respiration. The factors that affect the rate of cellular respiration are described in Table 3.2.4.

A word equation for cellular respiration is:

glucose + oxygen ⟶ carbon dioxide + water + energy

A balanced chemical equation for cellular respiration is:

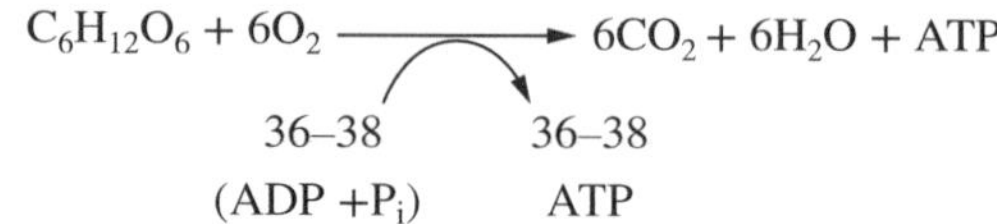

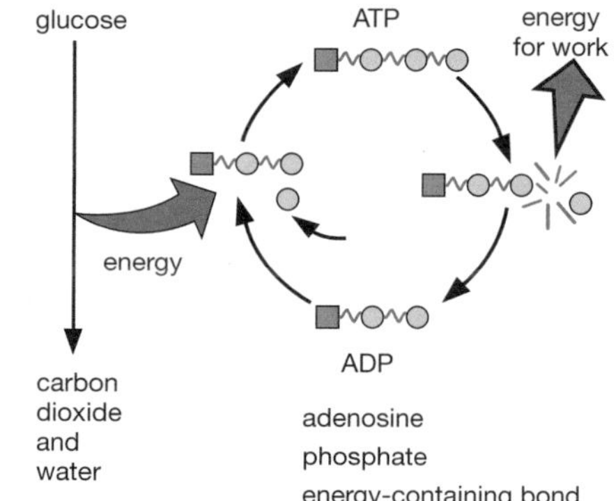

Figure 3.2.8 ATP/ADP cycle

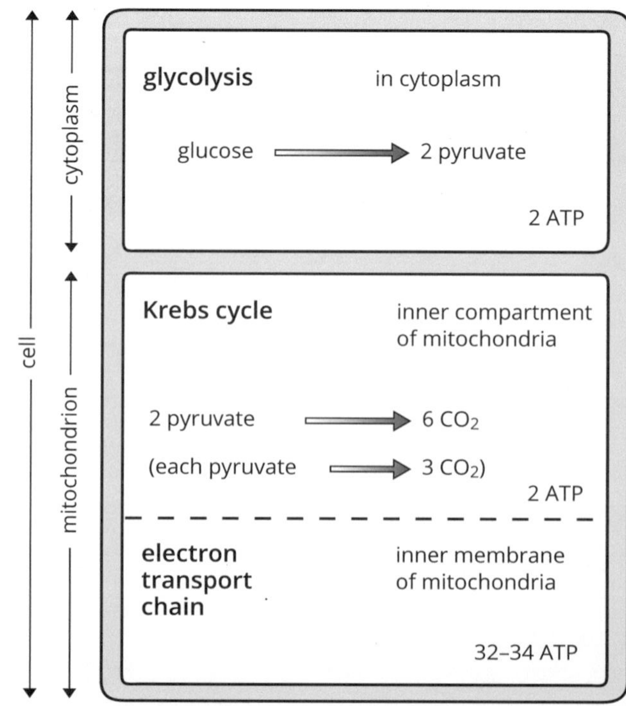

Figure 3.2.9 Energy release in the presence of oxygen—glycolysis, Krebs cycle and electron transport chain

ISBN 978 0 6557 0026 5

KEY KNOWLEDGE

Table 3.2.3 Cellular respiration—inputs and outputs

Site	Stage	Inputs	Outputs
cytoplasm	glycolysis	• 1 glucose molecule (6-C) • 2 NAD^+ • 2 ADP + 2 P_i	• 2 pyruvate molecules (3-C) • 2 NADH • 2 ATP
mitochondrion—inner compartment (fluid matrix)	Krebs cycle (citric acid cycle)	• 2 × Acetyl-CoA • 2 ADP + 2 P_i • electron acceptors NAD and FAD	• 6 × CO_2 (3 for each pyruvate) • 12 × H^+ (6 from each pyruvate) • loaded electron acceptors NADH and $FADH_2$ • 2 ATP
mitochondrion—folded inner membrane	electron transport chain	• NADH and $FADH_2$ (hydrogen ions carried by loaded electron carriers) • 6 × O_2 • 32–34 ADP + 32–34 P_i	• 6 × H_2O • 32–34 ATP • NAD and FAD (electron carriers are unloaded and available to be used again)

* NAD: nicotinamide adenine dinucleotide (electron carrier)
FAD: flavine adenine dinucleotide (electron carrier)
NADH: reduced NAD (hydrogen acceptor)
$FADH_2$: reduced FAD (hydrogen acceptor)

Table 3.2.4 Factors that affect the rate of cellular respiration

Factor	Effect
temperature	Enzymes involved in cellular respiration interact with their substrate molecules most efficiently at the optimal temperature for particular species. When temperatures are lower than optimal, the rate of collisions between enzymes and their substrates is lower and the rate of reaction slows. When temperatures are higher than optimal, enzymes can be denatured and cellular respiration ceases.
glucose availability	Cellular respiration is reliant on a continuous input of glucose. The higher the concentration of glucose, the higher the rate of cellular respiration until all available enzymes are engaged in the reaction. At this point the rate of reaction cannot increase and plateaus.
oxygen concentration	Oxygen is a requirement for cellular respiration. In the absence of oxygen, the electron transport chain cannot proceed and the energy needs of the cell rely on the anaerobic pathway to provide low-level, short-term energy. Anaerobic fermentation yields only 2 ATP molecules per glucose molecule entering the reaction. This is not enough to meet the ongoing needs of cells.

Cytochromes are proteins involved in the process of electron transport on the inner membrane of mitochondria during cellular respiration.

Since these processes involve the release of energy, they are called **exergonic reactions**. The release of energy is the result of the breakdown of organic compounds. Energy is released when chemical bonds are broken. These reactions are catabolic.

Anaerobic fermentation is a much less efficient process than the aerobic pathway (cellular respiration), but it allows organisms to have enough energy available to meet their needs, even when oxygen is not present. When working muscle cells are depleted in oxygen, the anaerobic pathway still meets basic energy needs. Figure 3.2.10 summarises this process. Some bacteria also respire anaerobically, with the chemical output varying depending on the species. *Lactobacillus* bacteria, active in the yoghurt-making process, respire anaerobically to produce lactic acid.

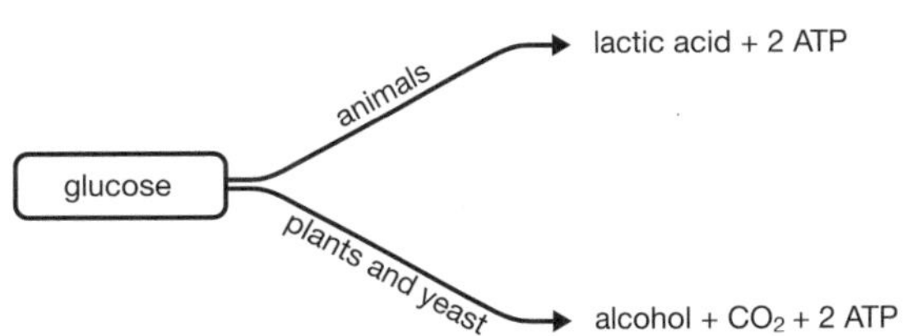

Figure 3.2.10 Energy release in the absence of oxygen

KEY KNOWLEDGE

Biotechnological applications of biochemical pathways

Cellular respiration and photosynthesis are biochemical pathways that are critical to life on Earth. Cellular respiration allows cells to release chemical energy to fuel cellular activity. Photosynthesis in green plants harnesses light energy from the Sun, essentially constructing those energy-rich organic molecules in the first place.

Our increasing understanding of these and other cell processes, including transcription and translation of DNA, provides a basis for research and development of biotechnologies that can be applied in medicine and agriculture to improve wellbeing. The development of the CRISPR-Cas9 gene-editing system is a significant recent example of such biotechnologies. It is a game-changer because it allows genes to be edited in living cells. Potential applications for this biotechnology include editing faulty genes to improve outcomes for people with genetic disorders, such as cystic fibrosis. Agricultural applications for crops and livestock include improved yield, increased nutritional content, decreased susceptibility to pests and disease and increased resilience to environmental factors such as drought. Canola that has been edited by CRISPR-Cas9 is already grown commercially in the United States. The implications for agriculture are profound, given the technology could be used to edit genes that regulate biochemical pathways such as photosynthesis, with a view to increased photosynthetic efficiency. See page 10 for further details and applications related to CRISPR-Cas9.

BIOMASS AND BIOFUELS

With Earth's human population nudging eight billion, the challenges that come with supporting the diverse needs of our vast global community are immense. Scientists are charged with finding the way forward. In this era of record atmospheric carbon dioxide concentration and climate change, the race is on to develop alternative ways of meeting food, space and energy needs. This involves removing carbon dioxide from the atmosphere and reducing the amount of greenhouse gases released in the first place. Fortunately, technologies already exist and are widely applied in many countries to mitigate issues around climate change. These include harvesting alternative energy sources such as solar, wind, geothermal and tidal energy.

Current research also centres on harvesting fuels from the **biomass** of particular living organisms to provide for our growing energy needs. Such renewable fuels are referred to as **biofuel**. The process involves the addition of yeasts or bacteria to carbohydrate-rich biomass, which is allowed to ferment before products such as ethanol and biogases are collected and purified. A specific focus on anaerobic fermentation means that alternative fuels can be available with reduced carbon dioxide production. In this way, energy needs can be met while at the same time reducing greenhouse gas emissions. Ethanol is already used in automotive fuel; biogas may be an alternative to natural gas.

 ISBN 978 0 6557 0026 5

Knowledge review—cell processes

This activity focuses on foundation ideas about cell process that you have studied before and on which the key ideas in this area of study are built. Key terms are reintroduced using their Latin or Greek origins. Understanding these origins can be valuable triggers that help to deduce the meanings of unfamiliar terms.

1 Use your knowledge and the information provided below to recall the meaning of each term. Provide examples or further explanatory notes where it is helpful to give extra context. Remember to be resourceful, checking glossaries, text material or a biology dictionary as needed.

	Scientific term	Meaning
a	photo	light
	synthesis	to make or produce
	photosynthesis	
b	chloro	green
	plastid	organelle that produces and stores organic molecules
	chloroplast	
c	respirare	to breathe in and out
	cellular respiration	
d	hydro	water
	glyco	sweet
	lysis	to split
	hydrolysis	
	glycolysis	
e	en	inside
	zyme	to leaven or make rise or transform
	enzyme	

2 Fermentation is another name for anaerobic respiration.

a Outline what this term means.

b Name a kind of cell or a kind of organism that you know can respire anaerobically.

ISBN 978 0 6557 0026 5

WORKSHEET 12

Enzymes as catalysts

1 Make a selection from the list below to fill in the missing words. This will give you a complete summary of enzymes and their functions.

biomolecules	deficiency	substrate-specific	optimal
denatured	catabolic	catalysts	temperature
protein	active site	photosynthesis	cellular respiration
metabolism	anabolic	coenzymes	enzyme-substrate complex
non-competitive	irreversible	regulatory	competitive

Enzymes are biological ________________. They increase the rate at which chemical reactions occur in living cells. They are organic compounds composed of ________________. Enzymes are important facilitators of energy-transforming reactions, recycling processes, synthesis of some compounds and breakdown of others—all of the processes that make up the ________________ of cells.

Enzymes are involved in both the construction and breakdown of ________________. Chemical reactions in which larger molecules are constructed from smaller ones are called ________________ reactions. ________________ is an example that typically occurs in plant cells. ________________ reactions involve the breakdown of large molecules into smaller ones. ________________ ________________ is an example of such a reaction in cells.

Enzymes are ________________-________________, that is, they have an active site that fits a particular substrate molecule only. This feature of enzymes is sometimes referred to as 'lock-and-key' or 'induced-fit'.

Enzymes have ________________ conditions; that is, they operate most efficiently under particular conditions.

Enzymes are sensitive to factors such as ________________ and pH. Their rate of activity can also be influenced by the concentration of substrate or enzyme. Cofactors and ________________ also affect enzyme action. The ________________ ________________ of an enzyme can be ________________ by excessive temperatures or pH.

Enzyme ________________ can result in disease.

________________ inhibitors are molecules that have a similar shape to the substrate. They compete for space in the active site of enzymes, preventing the formation of the ________________ ________________ ________________, thereby reducing the rate of enzyme-catalysed reactions. ________________ enzyme inhibitors are molecules that bind to other locations on enzymes, not the active site. They also have some impact on the shape of the active site, resulting in a decrease in enzyme efficiency.

Enzyme inhibition can be reversible or ________________. Reversible inhibition has an important ________________ role in cells.

2 Circle the word(s) that make the following statements about enzymes correct.

Enzymes are needed in small/large amounts.

Enzymes are used up/not used up in chemical reactions.

Enzymes can/cannot be used over and over again.

Enzymes do/do not affect the final amount of product in a reaction.

Enzymes increase/decrease the activation energy required to initiate reactions.

ISBN 978 0 6557 0026 5

WORKSHEET 13

Modelling

Features and function of enzymes

Enzyme action

Enzymes catalyse chemical reactions in cells in two directions, that is, the reactions are reversible. The diagram below illustrates an enzyme catalysing the production of the disaccharide (two sugar) sucrose molecules from the simple sugars glucose (G) and fructose (F).

1 Complete the diagram by inserting the glucose and fructose molecules into the active site of the enzyme. Then draw the newly constructed molecule of sucrose after it is released from the enzyme.

Use coloured pencils to colour-code your diagram. Use different colours for each of the different kinds of sugar molecule and for the enzyme. Label the enzyme–substrate complex.

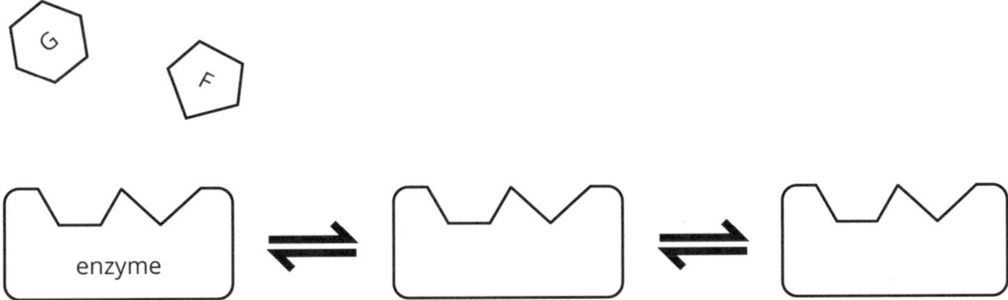

2 Explain why the relationship between enzymes and their substrates is described as 'lock-and-key'.

3 Name two factors that can affect the active site of an enzyme.

4 Describe what happens to the active site of an enzyme if it is denatured. What are the consequences for enzyme activity when this occurs?

5 Outline what is meant by the 'induced-fit' model. Use a diagram to illustrate.

Activation energy

Chemical reactions need a minimum energy input to 'kick-start' them. This is called the activation energy of the reaction. Enzymes reduce the amount of energy needed to provide this kick-start.

In the reaction in question 1, a more complex molecule is being constructed from two simpler ones. This involves a change in energy (ΔG) from an initial low level of energy to a final higher state. That is, there has been an overall input of energy into this reaction to construct the larger molecule. The energy that has been channelled into the reaction is contained within the chemical bonds that hold the two sugar subunits together. This reaction is called an endergonic reaction.

6 Graph A shows the activation energy for the sucrose-building reaction. The same enzyme catalyses the breakdown of sucrose into glucose and fructose in an exergonic reaction.

Complete Graph B to illustrate the activation energy for this reaction.

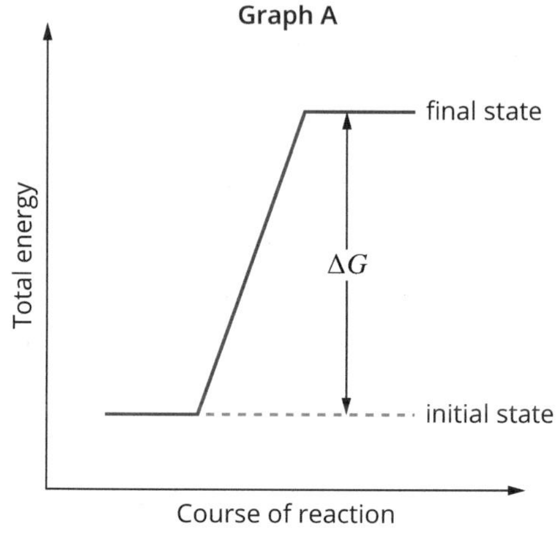

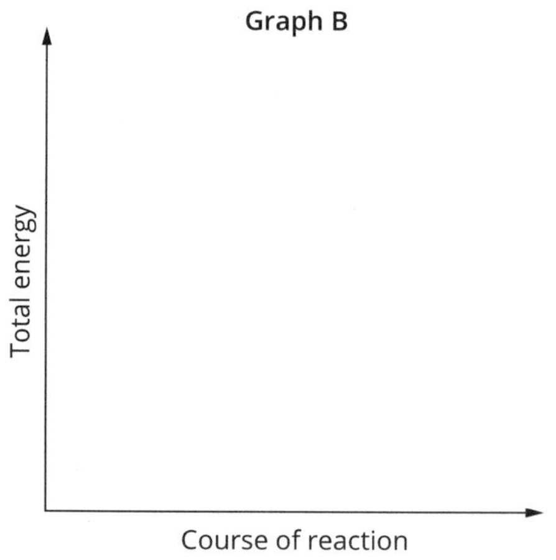

7 Why is the reaction in Graph B described as exergonic?

 ISBN 978 0 6557 0026 5

WORKSHEET 14

Chloroplasts and photosynthesis

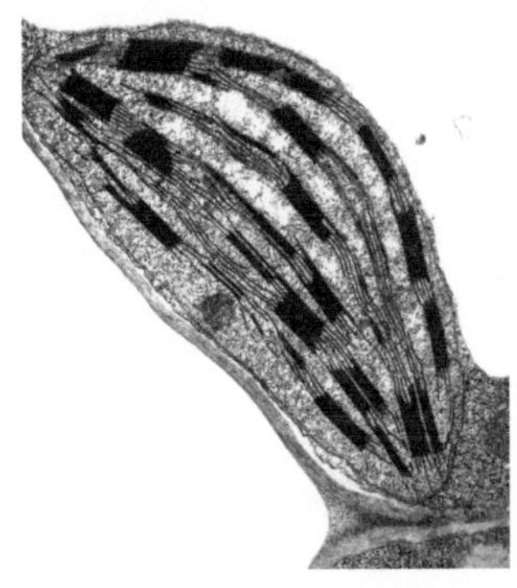

Figure 3.2.11 Transmission electron micrograph of chloroplast

In photosynthesis, green plants manufacture their own organic compounds using the raw materials of carbon dioxide and water. The process requires the input of light energy harnessed from sunlight. The pigment responsible for photosynthesis is chlorophyll. In green plants, the chlorophyll is contained in organelles called chloroplasts (Figure 3.2.11).

Photosynthesis occurs in two stages: the first in the presence of light, called the light-dependent reaction; the second in the absence of light, called the light-independent reaction.

1 Complete the sentences below that summarise key structures and reactions in the process of photosynthesis. For the chemical reactions, include specific information related to the numbers of molecules and ions.

Electron microscopy allows us to zoom in on the structure of chloroplasts. This technology reveals a complex network of stacked membranes that contain the chlorophyll. These membranous stacks are surrounded by a fluid matrix.

- The chlorophyll-filled membranous stacks are called the ________________.
- These membrane stacks are the site of the ________________-________________ reaction.
- The fluid matrix that surrounds these membranous stacks is called the ________________.
- The fluid matrix is the site of the ________________-________________ reaction.

2 Fill in the spaces to complete the ultra-structure of chloroplast diagram.

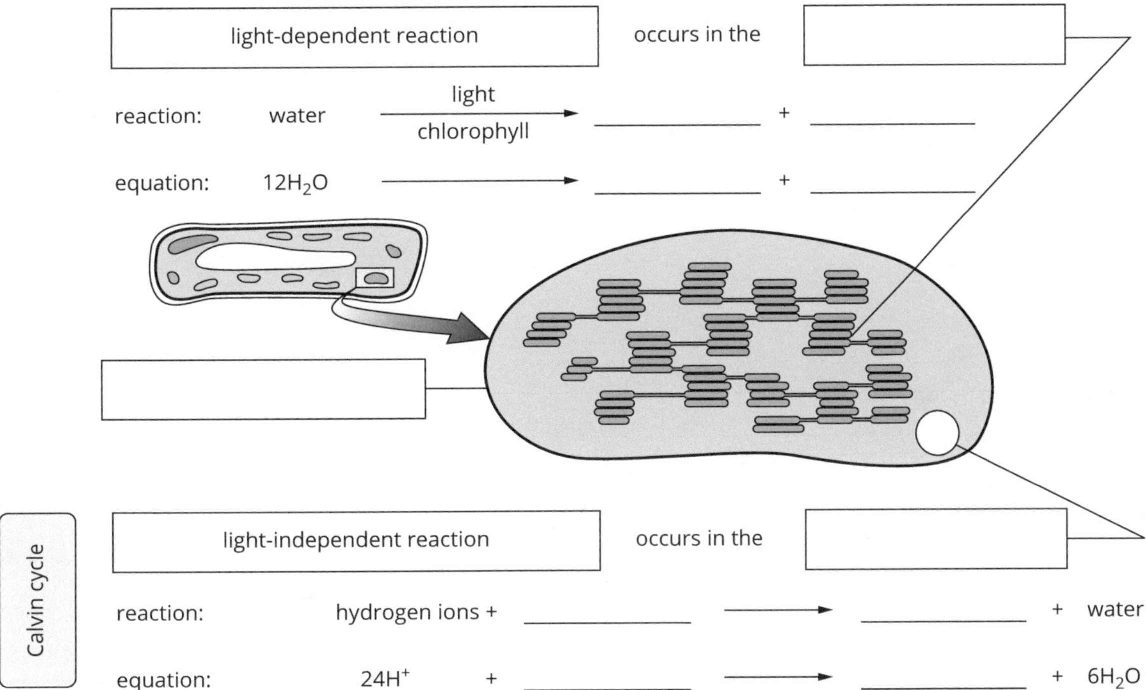

3 Name two factors that affect the rate of photosynthesis. Outline the effect in each case.

Factor 1: ____

____

Factor 2: ____

____

WORKSHEET 15

Classification and identification • Modelling

Mitochondria and cellular respiration

1 Define aerobic cellular respiration.

____

____

2 Write a balanced equation for cellular respiration in the space below.

3 Cellular respiration occurs in three stages. These are glycolysis, ____________ ____________ and ____________ ____________ ____________.

____________ occurs in the cytoplasm.

The Krebs cycle and the ____________ ____________ ____________ occur in the ____________.

4 Complete the pictograph of the cell and mitochondrion to build a summary of the different steps involved in the process of cellular respiration and the sites where each step occurs within the cell.

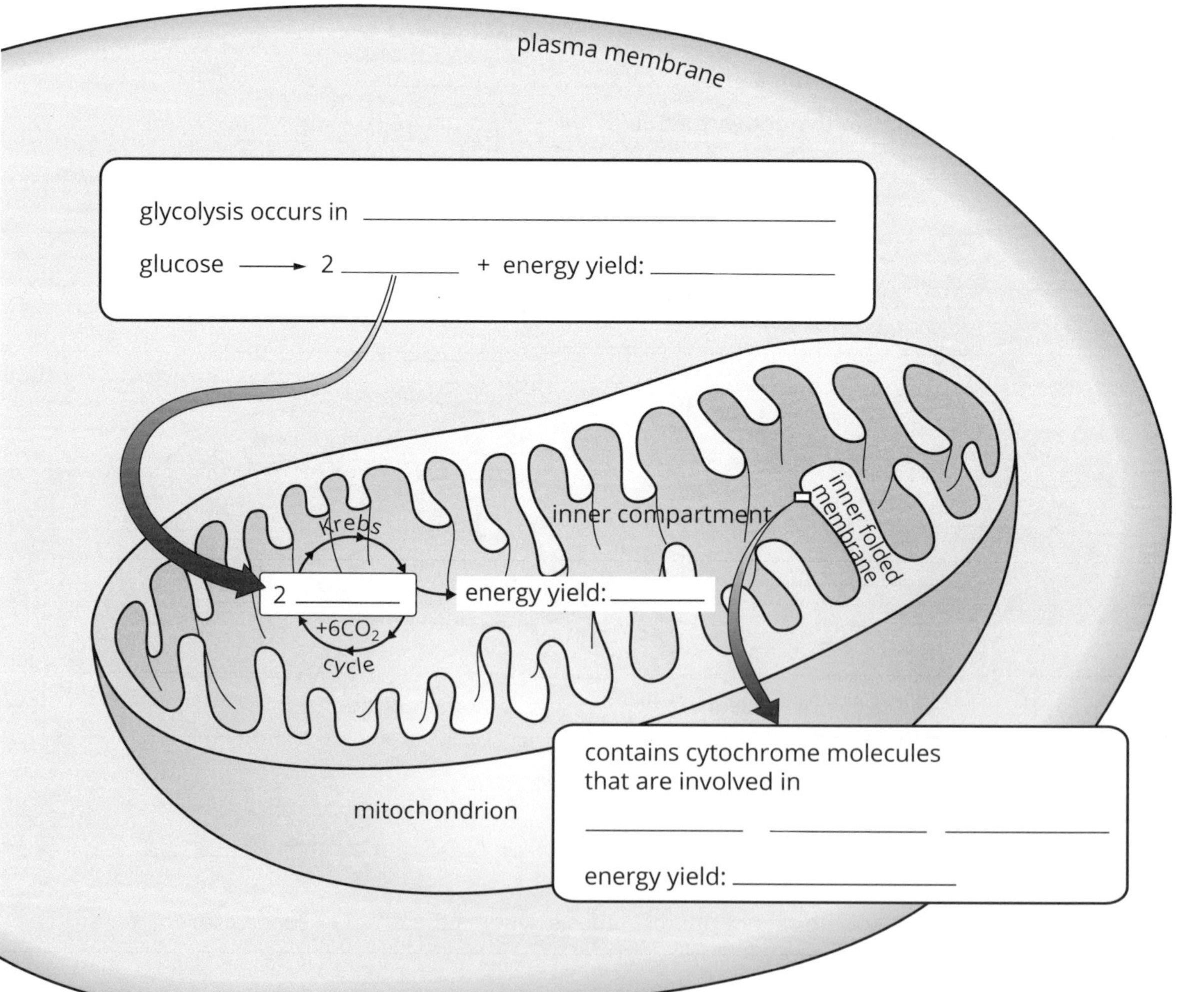

5 Mitochondria are sometimes referred to as the 'powerhouse' of cells. Suggest the reason for this.

____

____

 ISBN 978 0 6557 0026 5

WORKSHEET 16

Classification and identification • Modelling

Energy transformations in cells

1 Read the definitions listed in the boxes on the right side of the page. Choose the correct term from the list below to match each definition. Write each term in the box corresponding to its definition.

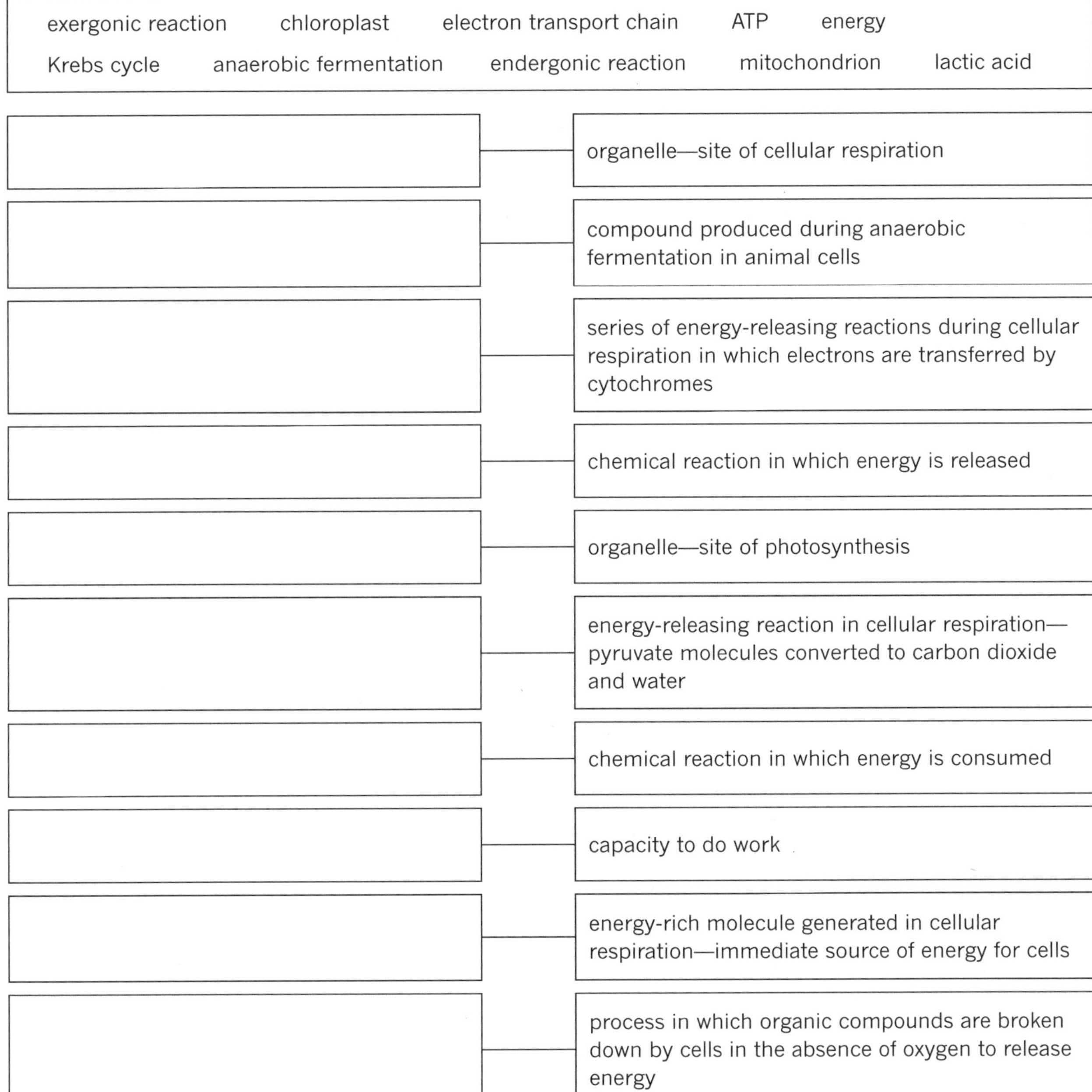

ISBN 978 0 6557 0026 5

2 Carefully examine each of the diagrams below. Write a couple of sentences about each to explain the process involved.

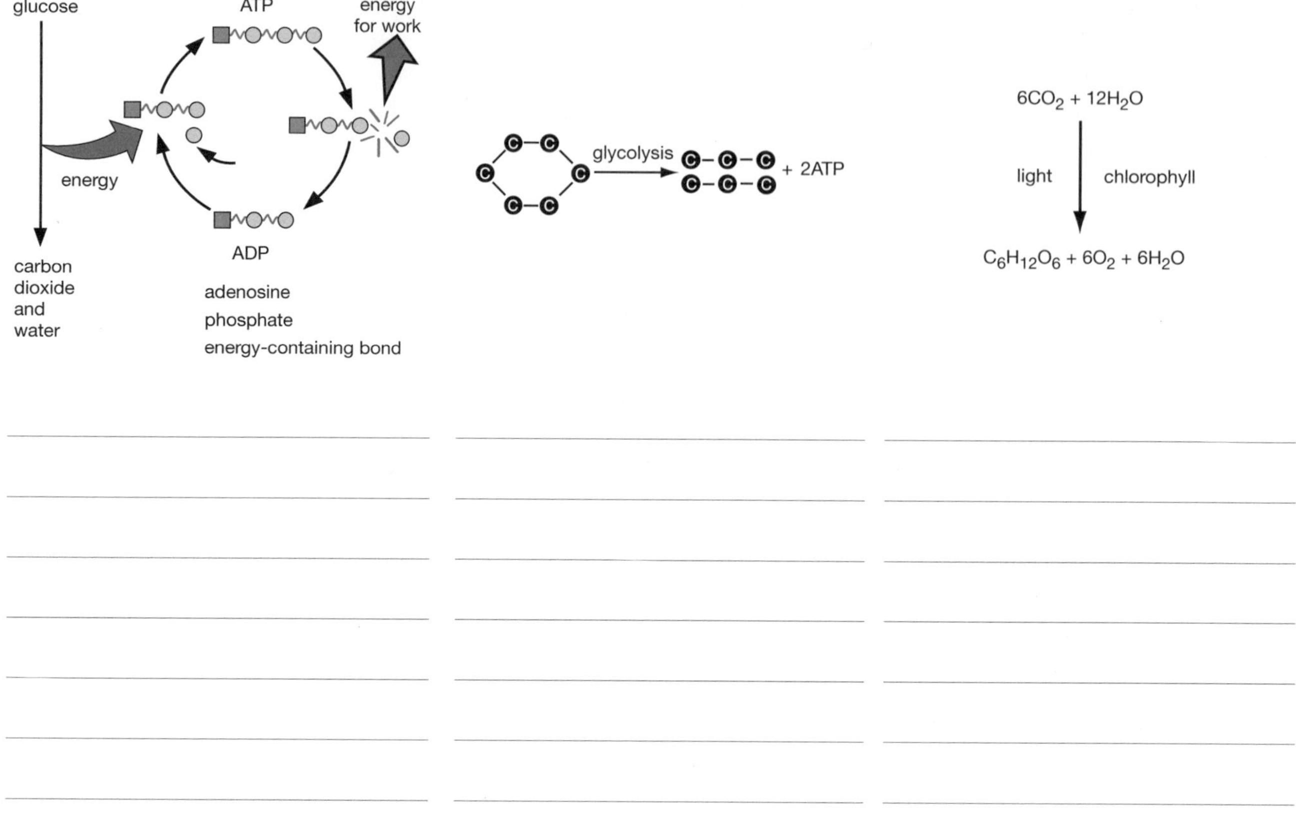

ISBN 978 0 6557 0026 5

WORKSHEET 17

Biomass for biofuels

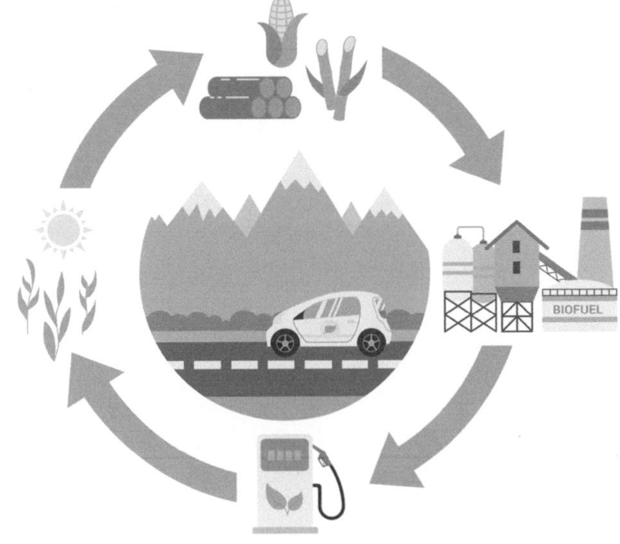

The production of biofuels from biomass is an industry on the rise in Australia and overseas. Biofuels can replace other sources of energy for the generation of electricity, gas for cooking and fuel for transport.

At least one biorefinery already operates in Australia, producing ethanol for the transport industry from the biomass of sorghum crops. Ethanol produced through the fermentation of sugars present in grain is added to petrol at a ratio of 10% ethanol : 90% petrol to produce the e10 or 91 petrol you see at the bowser.

Examine the flow chart in Figure 3.2.12 summarising sorghum biomass processed in biofuel production.

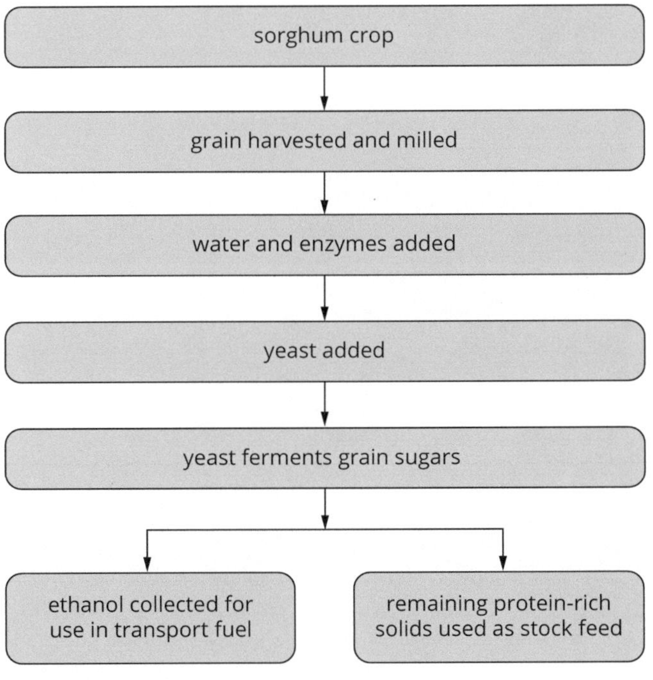

Figure 3.2.12 Sorghum biorefinery process

1 Define the term 'biofuel'.

2 Explain what is meant by fermentation, providing details of the inputs and outputs of fermentation in yeasts.

3 Use the information in Figure 3.2.12 and internet resources to identify two advantages of biofuels over fossil fuels.

4 Suggest a disadvantage of biofuels.

5 Identify two other examples of biofuels. For each, describe the application of the biofuel.

 ISBN 978 0 6557 0026 5

WORKSHEET 18

Reflection—How are biochemical pathways regulated?

The following table lists the key knowledge covered in this area of study.

1 Reflect on how well you understand the concepts listed. Rate your learning by shading the circle that corresponds to your current level of understanding for each one.

Key knowledge	Not confident ◄			► Very confident
Structure of the biochemical pathways in photosynthesis and cellular respiration	○	○	○	○
The role of enzymes and coenzymes in photosynthesis and cellular respiration	○	○	○	○
Factors that impact on enzyme function	○	○	○	○
Photosynthesis—inputs, outputs, locations of reactions and factors that affect the rate of this biochemical pathway	○	○	○	○
Cellular respiration—inputs, outputs, locations of reactions and factors that affect the rate of this biochemical pathway	○	○	○	○
Uses and applications of CRISPR-Cas9	○	○	○	○
Uses and applications of anaerobic fermentation	○	○	○	○

2 Consider the points you have shaded from Not confident to Very confident. List specific ideas you can identify that were challenging.

____

____

____

3 Write down two different strategies that you will apply to help further your understanding of these ideas.

____

____

____

PRACTICAL ACTIVITY 5

Controlled experiment

Factors affecting enzyme function

Suggested duration: 60 minutes

INTRODUCTION

Cellular respiration is the fundamental driving reaction that keeps cells alive and functioning, thereby maintaining the wellbeing of the overall organism. This is because it makes energy available to propel all other metabolic reactions. Biochemical pathways include reactions in which substances are constructed or digested. The by-products of some of these chemical reactions are harmful and must be removed from the body before they accumulate to a level at which they can cause tissue damage.

Carbon dioxide is one example of a metabolic waste produced by cells. It is carried to the lungs via the bloodstream, where it is removed from the body during exhalation.

Hydrogen peroxide is another waste product of cell metabolism. It is a potentially harmful chemical and must be removed immediately. The removal of hydrogen peroxide relies on the action of the enzyme catalase, which operates in cells to continually break down this waste material into harmless products. The efficiency of this enzyme depends on various factors, one of which you will look at in this investigation.

MATERIALS

- liver
- chopping board
- newspaper
- scalpel
- test tubes
- test-tube rack
- hydrogen peroxide
- Bunsen burner
- heat mat
- tripod
- gauze mat
- beaker of water
- mortar and pestle
- splint
- matches
- protective eyeglasses
- disposable gloves
- marker pen

AIM

- To design and conduct an investigation of enzyme activity.
- To consider factors that affect enzyme activity.

METHOD

Use the list of materials provided to design a controlled experiment to test whether or not:

- liver cells contain the enzyme catalase
- temperature affects enzyme activity.

Caution: Care is required when using sharp equipment such as a scalpel, and chemicals such as hydrogen peroxide. Check any safety procedures with your teacher.

ISBN 978 0 6557 0026 5

PRACTICAL ACTIVITY 5

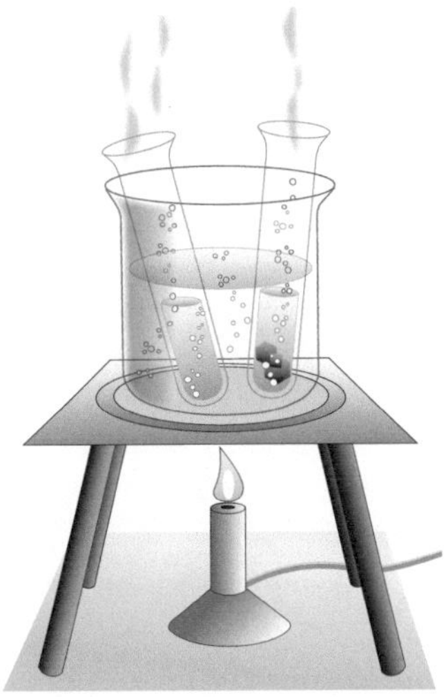

When you have completed your experimental design, check with your teacher before conducting your experiment.

Figure 3.2.13 Experimental set-up

1 Describe any evidence you observed that indicated enzyme activity occurred.

2 **a** How long did the activity you observed in the test tube last?

b What could you do to make the activity continue?

c What does this suggest about the way in which enzymes are involved in the chemical reaction?

3 The chemical equation below describes part of the reaction taking place.

$$2H_2O_2 \xrightarrow{\text{catalase}} 2H_2O + \underline{\qquad\qquad}$$

a Look carefully at the input into this reaction. Identify what gas is being produced.

b Suggest how you could test for the kind of gas produced. (Clue: look at the materials listed for this activity that you haven't used yet.)

4 Describe any differences you observe in the rate of enzyme activity for a small block of liver compared with the same amount of liver that has been processed using the mortar and pestle. Account for the differences you observe.

5 a Describe your observations when liver (block and ground) that has been boiled is exposed to hydrogen peroxide.

b Explain your observations.

CONCLUSIONS

6 Describe evidence from this investigation that supports the hypothesis that 'enzymes are not used up in the chemical reactions they catalyse and can be reused'.

7 Using your understanding of enzymes and referring to the experimental results achieved in this investigation, outline the effect of temperature on enzyme activity.

8 Referring to your experimental results, describe one other factor that affects enzyme activity.

 ISBN 978 0 6557 0026 5

PRACTICAL ACTIVITY 6

Controlled experiment

Investigating the activity of a selected plant enzyme

Suggested duration: 60 minutes

INTRODUCTION

The cells of living organisms are like tiny factories, each the centre of a great deal of activity including biochemical reactions that involve the breakdown of some molecules and the construction of others. Enzymes are the biological catalysts that facilitate these biochemical reactions. Without them, critical chemical reactions cannot occur or occur too slowly to maintain the wellbeing of cells, and the cells die. As well as cellular respiration and photosynthesis, many other life-sustaining biochemical pathways occur in cells. In plant cells, the carbohydrate-reducing enzyme diastase (an amylase) is an important enzyme involved in converting starch in storage organs to simple sugars that can be transported to growing tissue to be used in cellular respiration.

MATERIALS

- water baths set at 0°C, 20°C and 40°C
- 2 clean test tubes
- test-tube rack
- thermometer
- 2 × 10 mL measuring cylinders
- dropping pipettes
- white spotting tile with at least 10 wells
- iodine solution
- 1% starch solution—class stock
- 1% diastase solution—class stock
- toothpicks
- marker pen

AIM

- To investigate the activity of the plant enzyme diastase.
- To investigate the effect of temperature on enzyme activity.

BACKGROUND

Starch is a complex carbohydrate molecule made up of many glucose molecules (single sugars) joined together by chemical bonds. Starch stains a deep blue-black colour when mixed with iodine solution. Glucose and other simple sugars are not affected by iodine solution and so remain the brown colour of the iodine solution.

Diastase is a plant amylase, an enzyme that hydrolyses (breaks down in reaction with water) the chemical bonds between the glucose units in starch (Figure 3.2.14).

This activity makes efficient use of the time available when working in groups, with each group selecting a different temperature to test the diastase activity. Results for the different groups can then be shared and collated at the end of the experiment. Your teacher will help you organise your groups.

It will be useful to have the water baths set at the required temperatures prior to the commencement of class.

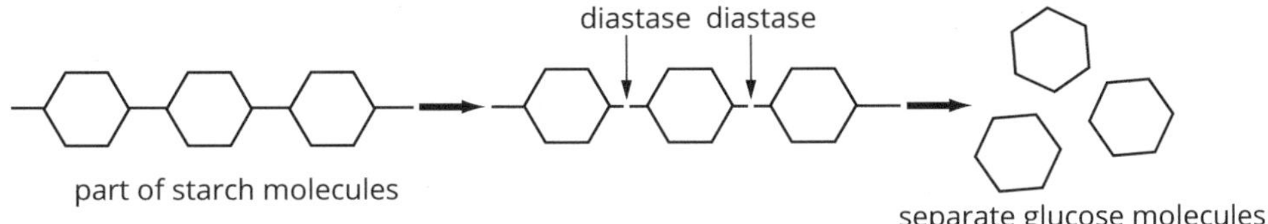

Figure 3.2.14 Effect of diastase on starch

METHOD

1. Use the first measuring cylinder to collect 5 mL of starch solution and pour it into one of the clean test tubes. Place the test tube in the test-tube rack. Use the second measuring cylinder to collect 5 mL of diastase solution and pour it into the other test tube. Place the test tube in the test-tube rack.
2. Check the temperature of the water in your water bath using the thermometer. Record the water temperature assigned to your group at the top of Table 2. If the water is at the required temperature, place the test-tube rack in the water bath. Allow five minutes for the solutions to adjust to the temperature of the water bath. While you are waiting, go on to the next step.
3. Use the marker pen to label the wells on the spotting tile. Mark the first two wells with the letters 'D' (diastase) and 'S' (starch) respectively. Then mark the remaining wells with the numbers 1 to 8. Place a drop of iodine solution into each well.

4 ▪ Use the dropping pipette to take a sample of the diastase solution. Place a drop into the first well. Record the colour observed in Table 1. Use a new pipette to repeat this procedure with the starch solution in the second well. Be sure to thoroughly rinse the pipette each time you use it.

5 ▪ After five minutes, quickly pour the contents of one test tube into the other, shake to mix and then return the test tube to the water bath. Note the time, and then use a new pipette to collect some of the starch/diastase mixture and place a drop into well number 1 on your spotting tile. Use a toothpick to mix the solutions in the well. Record your observations in Table 2.

6 ▪ After 60 seconds, take another drop of the starch/diastase mixture and place it in well number 2. Use a fresh toothpick to mix the solutions. Record your results. Repeat this procedure every minute until no further colour changes are observed.

7 ▪ Dispose of the solutions and equipment as directed by your teacher.

8 ▪ Use a spreadsheet to pool the class data. Enter the data from the different groups in your class into Table 3.

1 Suggest why it is important to test the starch solution and the diastase solution with the iodine at the start of this activity.

2 Why is it important to use a fresh pipette each time you test the starch/diastase mixture?

3 Suggest a hypothesis being tested in this activity.

4 Look carefully at the class data. Use your knowledge and understanding of enzymes to explain the results obtained at:

a 0°C

b 20°C

c 40°C

ISBN 978 0 6557 0026 5

PRACTICAL ACTIVITY 6

Table 1 Colour results

Solution	Iodine
starch	
diastase	

Table 2 Observation of colour changes at __°C

Colour changes for starch/diastase solution with iodine	
Time (min)	**Colour**
0	
1	
2	
3	
4	
5	
6	
7	
8	
9	
10	

Table 3 Class data

Group	Start	End
0°C		
20°C		
40°C		

CONCLUSIONS

5 Write a sentence that summarises the action of the enzyme diastase.

6 Referring to the experimental data in this activity, outline an environmental factor that appears to affect enzyme activity. Describe the evidence that supports your conclusion.

PRACTICAL ACTIVITY 7

Controlled experiment

Photosynthesis in variegated leaves

Suggested duration: PART A—lesson 1: experimental design 30 minutes; PART B—lesson 2 (several days later): set up experiment 30 minutes; PART C—lesson 3 (at least 24 hours later): check experimental results and answer questions 50 minutes

INTRODUCTION

Photosynthesis is the process in which green plants harness the Sun's light energy and use it to drive a reaction in which energy-rich carbohydrates are manufactured. Glucose molecules produced in photosynthesis can be stored as starch, a polymer of glucose. The set of reactions involved in photosynthesis occur in the chloroplasts, with the light-dependent stage occurring in the grana, which contain the pigment chlorophyll, and the light-independent stage subsequently taking place in the stroma. When light and chlorophyll are both present, photosynthesis can commence, combining carbon dioxide and water to produce glucose, with oxygen produced as a by-product.

In this activity, you will design a controlled experiment to investigate a given hypothesis that considers conditions required for photosynthesis to proceed. The background information below will support your experimental design by providing a method for testing for the presence of starch in leaves that have been exposed to light.

MATERIALS

- pot plant with variegated leaves
- light source
- 1 cm-wide strips of aluminium foil
- paperclips
- large test tubes
- test-tube rack
- electric hot plate
- methylated spirits
- 600 mL beaker
- evaporating dishes
- iodine/potassium iodine solution
- starch solution
- distilled water
- spotting tile
- 2 pipettes
- forceps

AIM

- To consider the relationship between chlorophyll distribution and starch production in variegated leaves.
- To investigate factors affecting photosynthesis in variegated leaves.

HYPOTHESIS

- Green leaf tissue produces starch when light is present.

BACKGROUND

Testing for the presence of starch in leaves

The presence of starch in leaves can be tested using the following method:

- Boil a leaf in water for three minutes.
- Remove the leaf using forceps and place it in a large test tube half-filled with methylated spirits. Immerse the leaf fully.
- Place the test tube in a beaker of water on an electric hot plate. Allow to boil for 10 minutes. The boiling methylated spirits will take on a green colour. The chlorophyll in the leaf dissolves and the leaf becomes uniformly pale (Figure 3.2.15).
- Remove the leaf from the test tube using forceps and wash it under cool, running water.
- Float the leaf in an evaporating dish containing iodine solution. Leave for three minutes. Iodine is a brownish-yellow solution that turns blue-black in the presence of starch.
- Remove the leaf from the evaporating dish using forceps and wash it under cool, running water.
- Float the leaf in an evaporating dish containing clean water.
- Observe the leaf for the distribution of blue-black patches.

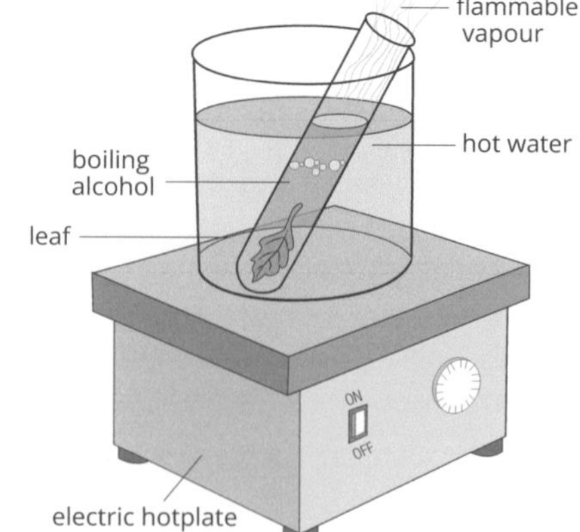

Figure 3.2.15 Extracting the chlorophyll

De-starching leaves

Storing a pot plant in a dark cupboard for several days is effective in ensuring starch has been depleted from leaves.

ISBN 978 0 6557 0026 5

PART A • EXPERIMENTAL DESIGN

1 Collect a clean spotting tile, starch, iodine and distilled water. Place a few drops of distilled water into one well on the spotting tile and a few drops of starch into another. Now add a drop of iodine to both. Record any colour changes observed.

Colour changes with addition of iodine		
	Distilled water	Starch solution
Iodine		

2 Design an experimental method that could be used to test the hypothesis set out above. Use the background information, as well as the materials listed, as a guide. Be sure to check your experimental method with your teacher before commencing the experiment.

You will need to prepare a risk assessment. Toolkit pages xi–xii will be helpful here. Include before and after sketches of leaves used in your experiment.

Experimental method:

PART B • EXPERIMENTAL SET-UP

Follow your experimental procedure to set up your experiment. (Clue: your pot plant will have been stored in the dark for several days by now.)

3 Suggest why storing a pot plant in the dark for several days contributes to the depletion of starch in leaves.

4 For the experiment you have conducted, identify the:

a independent variable

b dependent variable

5 List four factors you have kept constant in this experiment.

PART C • EXPERIMENTAL RESULTS (24 HOURS LATER)

6 Describe the reaction that occurs when starch is treated with iodine.

7 Draw a leaf showing the distribution of chlorophyll before the experiment. Then draw the same leaf showing the distribution of iodine after the experiment.

Leaf before

Leaf after

8 Explain the relationship between the distribution of chlorophyll and the distribution of starch in question 7.

9 Suggest steps that could be taken to improve the reliability of your experimental results.

CONCLUSION

10 Draw on your experimental results to explain whether the hypothesis is supported or refuted.

ISBN 978 0 6557 0026 5

PRACTICAL ACTIVITY 8

Controlled experiment

Photosynthesis and cellular respiration in a pondweed

Suggested duration: 20 minutes to set up; 50 minutes second lesson

BACKGROUND

Photosynthesis and cellular respiration are considered reciprocal reactions in the biochemical pathways of plants—the products of one reaction become the raw materials for the other. In photosynthesis, green plants combine carbon dioxide and water in the presence of light and chlorophyll, producing glucose and oxygen. During cellular respiration, glucose and oxygen are consumed to release energy for cells and carbon dioxide and water are produced.

When carbon dioxide is dissolved in water a chemical reaction takes place that results in the production of carbonic acid. The reaction can be summarised as follows:

carbon dioxide + water → carbonic acid

$$CO_2 + H_2O \rightarrow H_2CO_3$$

A simple test for the presence of dissolved carbon dioxide in water involves the indicator bromothymol blue. Bromothymol blue is sensitive to changes in pH, turning blue in alkaline conditions and yellow in acidic conditions.

MATERIALS

- *Elodea*
- 2 test-tube racks
- 4 test tubes
- 4 rubber stoppers
- pond water
- marker pen
- bromothymol blue indicator solution
- spotting tile
- distilled water
- ammonia
- dilute hydrochloric acid
- 24-hour light source

Elodea is a common pondweed that uses carbon dioxide in photosynthesis and generates carbon dioxide in cellular respiration. As *Elodea* is an aquatic plant, gaseous exchange occurs directly between the tissues of the pondweed and its watery environment. In this experiment you will test for the presence of carbon dioxide in different conditions.

AIM

- To investigate factors that affect the process of photosynthesis in an aquatic plant.
- To consider the relationship between the processes of photosynthesis and cellular respiration.

METHOD

Pre-test

1. Add a couple of drops of dilute hydrochloric acid to one well on your spotting tile.
2. Add a drop of bromothymol blue and record the colour in Table 1.
3. Repeat steps 1 and 2 using distilled water and ammonia.

Table 1 Colour change with bromothymol blue

Solution	Colour
dilute hydrochloric acid	
distilled water	
ammonia	

Experiment

4 ▪ Set up two test-tube racks, each with two test tubes. Use the marker pen to clearly label the test tubes A, B, C and D, as shown in Figure 3.2.16.

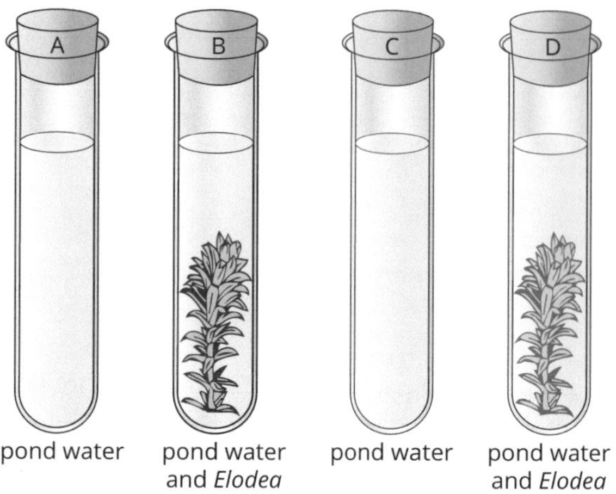

Figure 3.2.16 The experimental set-up and labelling of test tubes A through D

5 ▪ Fill each of the test tubes with pond water until they are almost full. Add 20 drops of bromothymol blue to each of the four test tubes.

6 ▪ Place a long piece of *Elodea* into each of test tubes B and D. Make sure the two pieces are of approximately equal length.

7 ▪ Tightly stopper each of the test tubes.

8 ▪ Record the colour of the solution in each test tube in Table 2.

9 ▪ Place test tubes A and B under a 24-hour light source and test tubes C and D in the dark.

10 ▪ Leave the experiment set up for a minimum of 24 hours.

Table 2 Results for indicator colour

	Test tubes in the light		Test tubes in the dark	
	A	B	C	D
Initial colour				
Final colour				

Next lesson (at least a full day later)

11 ▪ Check each of the test tubes and record the colour of solution for each in Table 2.

DISCUSSION

1 Write a hypothesis for this investigation.

2 a Which test tubes represent the control(s) in this experiment?

b Outline the purpose of the control.

3 Describe two ways in which you controlled this experiment.

 ISBN 978 0 6557 0026 5

4 What does a change in colour suggest about the pH of the solution?

5 Which test tube(s) in this experiment showed a change in colour? Explain why this occurred.

6 Write out the balanced equation for photosynthesis.

7 Write out the balanced equation for cellular respiration.

CONCLUSIONS

8 a State one conclusion you can draw from the experiment set up in the light.

b Outline the evidence that supports this conclusion.

9 a State one conclusion you can draw from the experiment set up in the dark.

b Outline the evidence that supports this conclusion.

10 Summarise the relationship between the biochemical pathways of photosynthesis and cellular respiration in *Elodea*.

EXAM QUESTIONS

Multiple-choice questions

Question 1 VCE Biology 2017 (A) 5

The biochemical pathway of glycolysis involves nine intermediate reaction steps.

One of these steps is represented in the diagram below.

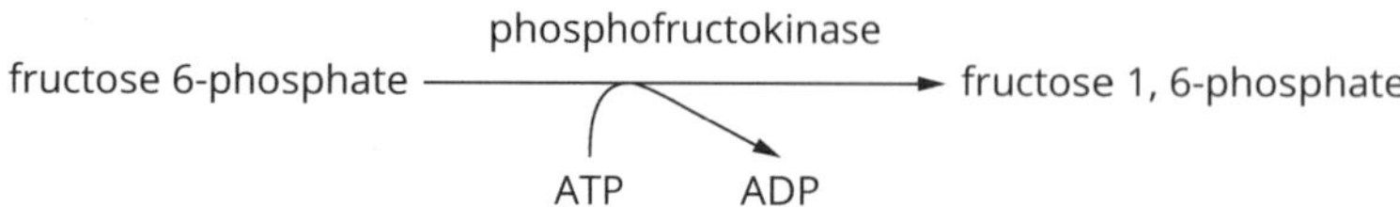

It is correct to state that, in this reaction, phosphofructokinase

A. acts as a coenzyme.

B. increases the rate of reaction.

C. is the substrate for the reaction.

D. releases energy in the form of ADP.

Use the following information to answer questions 2–4. VCE Biology 2017 (A) 6-8

Hydrogen peroxide is a toxic by-product of many biochemical reactions. Cells break down hydrogen peroxide into water and oxygen gas with the help of the intracellular enzyme catalase. The optimum pH of catalase is 7.

A Biology student measured the activity of catalase by recording the volume of oxygen gas produced from the decomposition of hydrogen peroxide when a catalase suspension was added to it. The catalase suspension was made from ground, raw potato mixed with distilled water. The student performed two tests and graphed the results.

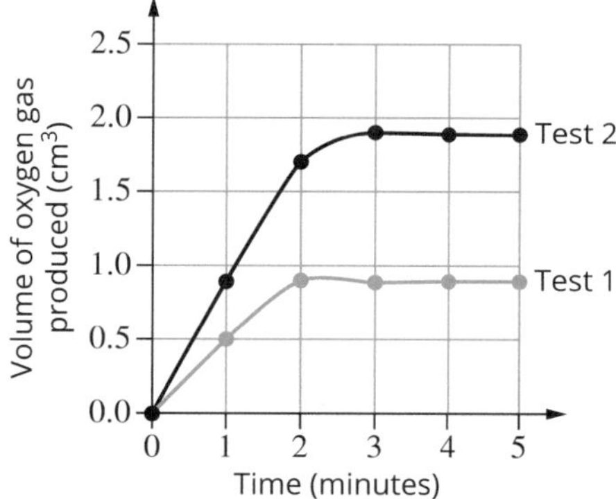

Test 1 used 5 mL of 3% hydrogen peroxide solution and 0.5 mL of catalase suspension, and was conducted at 20°C in a buffer solution of pH 7.

Test 2 was carried out under identical conditions to Test 1, except for one factor that the student changed.

Question 2 VCE Biology 2017 (A) 6

An explanation for the results of Test 2 would be that the student

A. increased the concentration of catalase by adding less water to the ground potato.

B. increased the temperature by placing the test tube in a water bath set at 30°C.

C. used a hydrogen peroxide solution with a higher concentration.

D. added a catalase suspension made from a cooked potato chip.

 ISBN 978 0 6557 0026 5

EXAM QUESTIONS

Question 3 VCE Biology 2017 (A) 7

The student then performed more tests by varying the pH of the buffer solution.
It is expected that

A. at pH 6 the reaction will cease.

B. at pH 9 the reaction will be faster.

C. at pH 2 and pH 10 very little oxygen will be produced.

D. a greater volume of oxygen will be produced each time the pH is increased.

Question 4 VCE Biology 2017 (A) 8

During the experiment, the student measured the varying pH levels using a digital pH meter. The student calibrated the meter using a pH 7 buffer solution.

The reason the student calibrated the pH meter was to

A. ensure a random error would not influence the results.

B. eliminate the effect of all uncontrolled variables.

C. enable the use of the instrument with precision.

D. allow the pH to be measured accurately.

Use the following information to answer questions 5 and 6. VCE Biology 2017 (A) 13 and 14

The graph below shows the net output of oxygen in spinach leaves as light intensity is increased. Temperature is kept constant during the experiment.

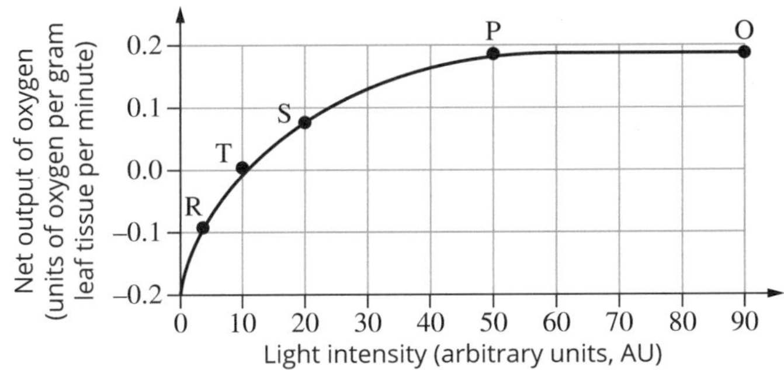

Question 5 VCE Biology 2017 (A) 13

Which one of the following conclusions can be made based on the graph?

A. At point T photosynthesis is no longer occurring.

B. The optimal level of light intensity for photosynthesis is 40 AU.

C. At point S the amount of oxygen output is a third of that at point P.

D. Below 10 AU of light intensity the aerobic respiration rate is greater than the photosynthesis rate.

Question 6 VCE Biology 2017 (A) 14

The rate of oxygen output remains constant between points P and O because

A. heat has denatured the enzymes involved in the photosynthesis reactions.

B. the concentration of available carbon dioxide limits the rate of photosynthesis.

C. the light intensity has damaged the chlorophyll molecules present in the spinach chloroplasts.

D. high levels of oxygen produced at point P have accumulated around the spinach leaves, resulting in no more oxygen being produced.

EXAM QUESTIONS

Question 7 VCE Biology 2018 (A) 8

The diagram below shows a section through a part of a mitochondrion.

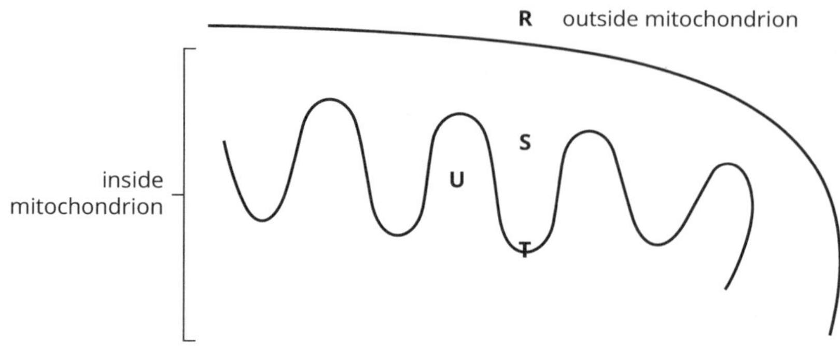

The sites of the pathways in aerobic respiration are

A. R—glycolysis, S—Krebs cycle, T—electron transport chain.

B. U—glycolysis, T—Krebs cycle, R—electron transport chain.

C. R—glycolysis, U—Krebs cycle, T—electron transport chain.

D. T—glycolysis, R—Krebs cycle, S—electron transport chain.

Question 8 VCE Biology 2018 (A) 9

Which of the following gives the inputs and outputs of the electron transport chain in an animal cell?

	Inputs	Outputs
A.	NADH, ADP, oxygen, P_i	ATP, NAD^+, water
B.	NADH, ADP, water, P_i	ATP, NAD^+, oxygen
C.	NAD^+, ADP, oxygen, P_i	NADH, ATP, water
D.	NADPH, ADP, water, P_i	$NADP^+$, ATP, oxygen

Question 9 VCE Biology 2018 (A) 10

Which pair of molecules contains the greatest amount of stored energy?

A. NADH and ATP

B. NAD^+ and ATP

C. NAD^+ and ADP

D. NADH and ADP

Question 10

Biorefineries are on the rise, replacing traditional fossil fuels as a source of energy to meet the demands of society, both in homes and industry. Biorefineries use organisms such as bacteria and yeast to ferment the biomass of particular organisms, harvesting products such as ethanol as a petrol alternative and biogas as a natural gas alternative. By-products of biorefining may also have value for other applications, for example, as stock feed.

An advantage of biorefineries as a means of acquiring energy and fuel is

A. it is an efficient means of aerobic respiration, maximising energy output.

B. they can use any kind of biomass as a substrate for fermentation.

C. it produces less carbon dioxide than fossil fuels, thereby helping to mitigate the impacts of climate change.

D. it releases excess oxygen into the atmosphere, thereby combatting the effects of greenhouse gas accumulation.

 ISBN 978 0 6557 0026 5

EXAM QUESTIONS

Short-answer questions

Question 1 (8 marks) VCE Biology 2016 (B) 2

Plant materials containing cellulose and other polysaccharides are reacted with acids to break them down to produce glucose. This glucose is then used by yeast cells for fermentation.

a. Why is fermentation important for yeast cells? 1 mark

b. What are the products of fermentation in yeast cells? 1 mark

A by-product of the acid treatment of plant materials is a group of chemical compounds called furans. It has been observed that as the concentration of furans increases, the rate of fermentation decreases. The enzyme alcohol dehydrogenase is required for the process of fermentation.

c. Design an experiment to test the hypothesis that one of the furans, called furfural, is an inhibitor of the enzyme alcohol dehydrogenase. Assume that the experiment will be repeated many times and that environmental factors are kept constant. 4 marks

d. Scientists have proposed that furfural is a competitive inhibitor of the enzyme alcohol dehydrogenase. 2 marks

Explain how furfural could act as a competitive inhibitor of the enzyme alcohol dehydrogenase.

EXAM QUESTIONS

Question 2 (8 marks) VCE Biology 2017 (B) 11

Matthew investigated how changes in environmental temperature affected oxygen (O_2) and carbon dioxide (CO_2) levels in the air around a cockroach. He used three digital probes linked to a computer, a closed animal chamber and a heat lamp in the experimental set-up shown.

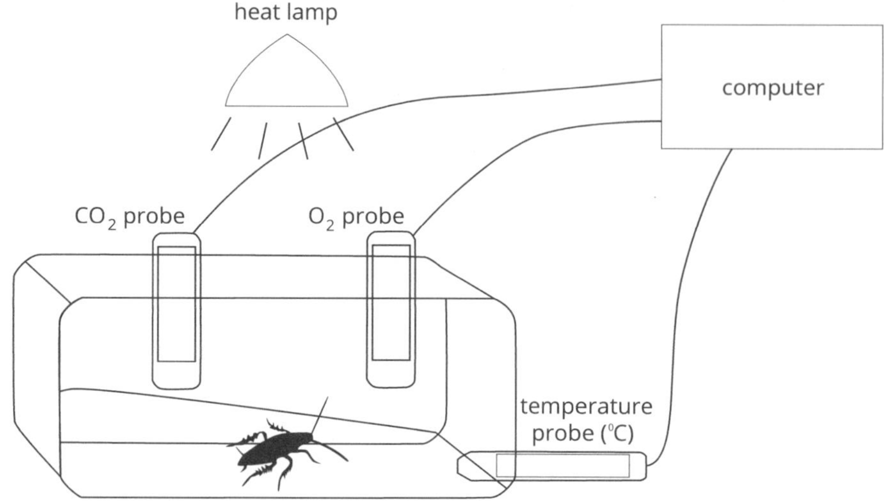

a. Name the cellular process being investigated in Matthew's experiment. 1 mark

____

b. Identify the 2 marks

- dependent variables

____

____

- independent variable

____

Before placing the cockroach in the chamber, Matthew decided to measure the temperature, and carbon dioxide and oxygen levels for four minutes. The following results were recorded.

Time (minutes)	CO_2 (%)	O_2 (%)	Temperature (°C)
0	0.04	22.3	29.5
1	0.04	22.1	29.8
2	0.04	22.0	30.0
3	0.04	22.0	30.0
4	0.04	22.0	30.0

c. Explain why Matthew recorded the data for four minutes and not just one minute. 1 mark

____

____

____

 ISBN 978 0 6557 0026 5

EXAM QUESTIONS

After the initial four-minute period, Matthew quickly placed the cockroach in the chamber and began recording the data from the digital probes. After 10 minutes, he placed ice packs around the sides of the animal chamber to slowly bring the temperature of the chamber down to 10 °C. He recorded the data using the digital probes for a further 20 minutes. He repeated the experiment once every day for the next six days with the same cockroach. At all times, he took care to ensure that the cockroach showed no signs of stress.

d. Other than repeating the entire experiment, identify **two** control measures Matthew should have included in his experimental design. Explain how each of these control measures could affect the results if not kept constant. 4 marks

Matthew constructed the following graphs from the averaged results of the seven experiments.

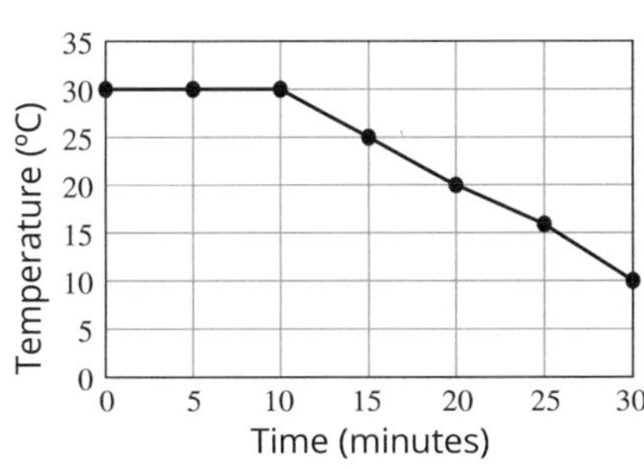

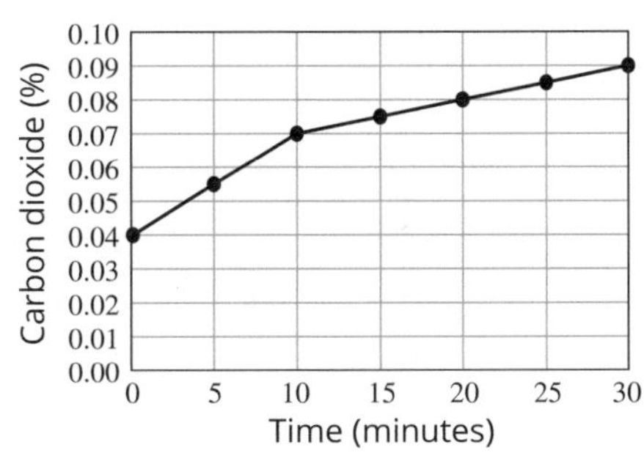

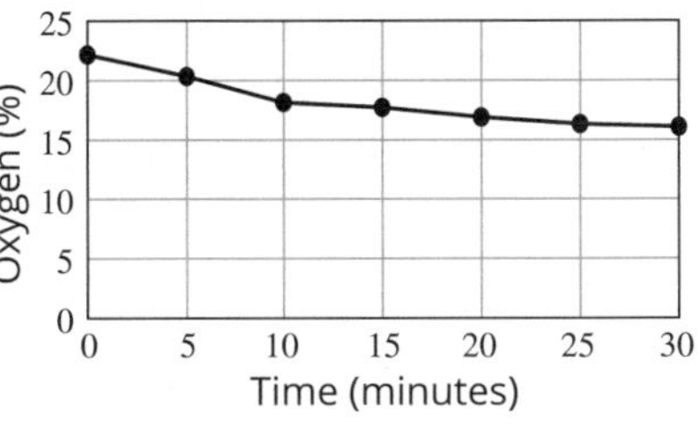

e. i. Using the graphical data, describe the changes in the levels of carbon dioxide and oxygen when the temperature in the chamber was kept constant compared to when the temperature was decreasing. 4 marks

ii. What conclusion do you think Matthew can draw from his investigation? You should refer to each of the following in your response: 4 marks

- the cellular process named in **part a**.
- the variables identified in **part b**.
- the evidence collected during Matthew's experiments.

Question 3 (4 marks) VCE Biology 2015 (B) 3

Below is a diagram of a chloroplast.

X

a. Name the structure labelled X. 1 mark

b. Complete the following table by referring to the diagram above and your knowledge of photosynthesis. 3 marks

Name of the stage of photosynthesis that occurs at X		
Two input molecules that are required for reactions at X	1.	2.
Two output molecules that result from the reactions at X	1.	2.

 ISBN 978 0 6557 0026 5

EXAM QUESTIONS

Question 4 (11 marks) VCE Biology 2018 (B) 11

Elsa read that red algae survive at greater water depths than green algae because of a pigment in the red algae called phycoerythrin. This pigment enables the algae to absorb more of the green light available at greater water depths. Elsa decided to investigate this by carrying out an experiment.

Using a standard technique, the single-celled algae were trapped in jelly balls. One set of balls contained green algae and another set contained red algae.

To measure the rate of photosynthesis, Elsa used a stopwatch and the pH indicator phenol red. Phenol red changes colour in solutions with different concentrations of carbon dioxide. In low carbon dioxide concentrations, phenol red is pink and in higher carbon dioxide concentrations it is yellow.

Elsa placed the jelly balls into test tubes and covered them with a solution containing dissolved carbon dioxide. Phenol red indicator was added to each solution.

The diagram below shows the set-up of Elsa's experiment.

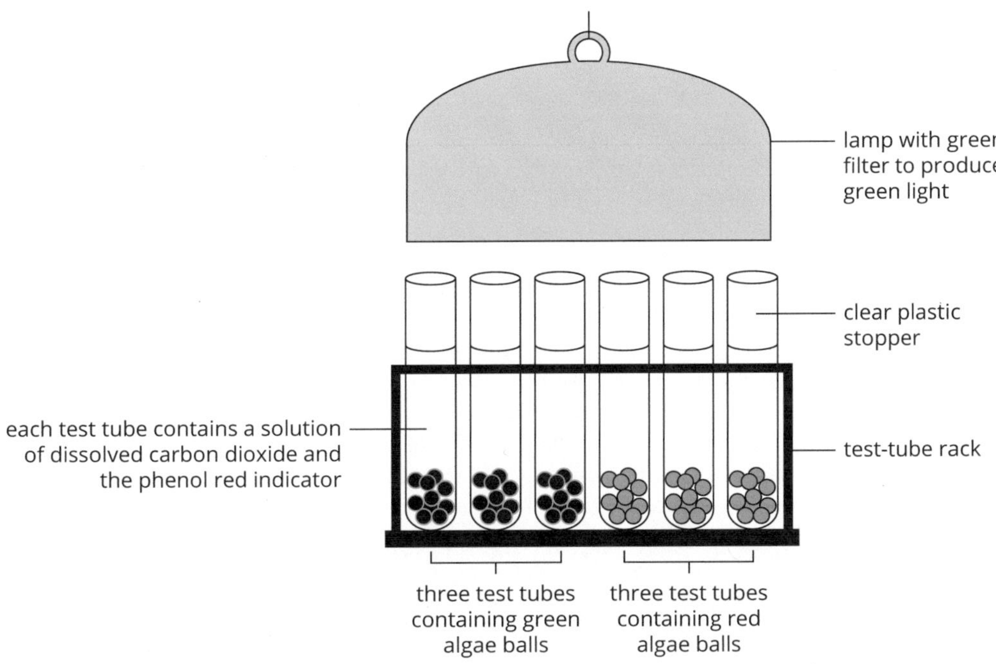

a. State the hypothesis that Elsa was testing. 1 mark

____

____

b. List three variables that would need to be controlled to ensure the experiment produced valid results. 3 marks

1. ______________________________

2. ______________________________

3. ______________________________

c. State the independent variable and the dependent variable in this experiment. 2 marks

Independent variable: ______________________________

Dependent variable: ______________________________

d. What results would disprove the hypothesis of Elsa's experiment? 2 marks

e. Elsa's laboratory partner suggested that they should also set up an identical experiment but keep the test tubes and their contents in the dark. 1 mark

Explain why this is a good suggestion.

f. Elsa's teacher said that even if the students completed the additional experiment in the dark and even if all of those results supported the hypothesis, there may still be other plausible explanations for their results. 2 marks

Suggest **two** other explanations to which the teacher could be referring.

Question 5 (6 marks)

CRISPR stands for 'clustered regularly interspaced short palindromic repeats'. It refers to a sequence of bacterial DNA composed of repeated palindromic nucleotide sequences that are interrupted or interspaced by unique DNA sequences. CRISPR RNA (crRNA) has been engineered to work in concert with a protein called Cas9, which has the property of being able to cut target DNA. Together, the CRISPR-Cas9 system allows scientists to edit DNA in living cells.

(A palindrome is a sequence of items that reads the same backwards and forwards.)

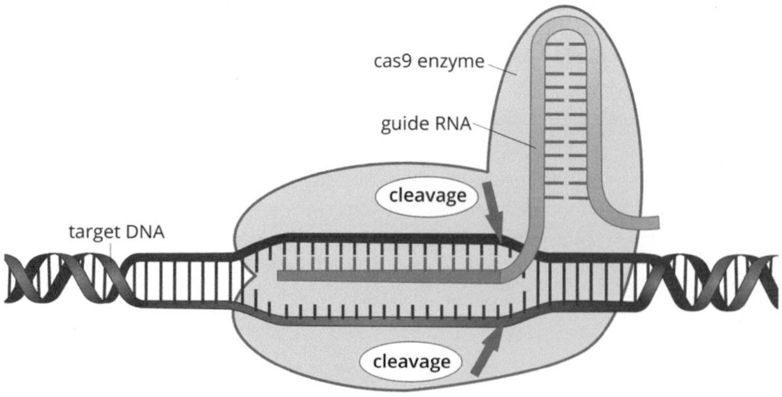

a. Scientists discovered that the unique spacer DNA sequences present in CRISPR arrays matches particular viral DNA sequences. 1 mark

Suggest the significance of this discovery in terms of our understanding of bacteria.

 ISBN 978 0 6557 0026 5

EXAM QUESTIONS

b. Describe the role of guide RNA in the CRISPR-Cas9 system. 1 mark

c. State one advantage of using CRISPR-Cas9 technology. 1 mark

d. State one disadvantage of using CRISPR-Cas9 technology. 1 mark

e. Identify two issues raised by the use of CRISPR-Cas9 technology. 2 marks

UNIT

4 How does life change and respond to challenges?

AREA OF STUDY 1

How do organisms respond to pathogens?

Outcome 1

On completion of this unit the student should be able to analyse the immune response to specific antigens, compare the different ways that immunity may be acquired and evaluate challenges and strategies in the treatment of disease.

Key knowledge

Responding to antigens

- physical, chemical and microbiota barriers as preventative mechanisms of pathogenic infection in animals and plants
- the innate immune response including the steps in an inflammatory response and the characteristics and roles of macrophages, neutrophils, dendritic cells, eosinophils, natural killer cells, mast cells, complement proteins and interferons
- initiation of an immune response, including antigen presentation, the distinction between self-antigens and non-self antigens, cellular and non-cellular pathogens and allergens

Acquiring immunity

- the role of the lymphatic system in the immune response as a transport network and the role of lymph nodes as sites for antigen recognition by T and B lymphocytes
- the characteristics and roles of the components of the adaptive immune response against both extracellular and intracellular threats, including the actions of B lymphocytes and their antibodies, helper T and cytotoxic T cells
- the difference between natural and artificial immunity and active and passive strategies for acquiring immunity

Disease challenges and strategies

- the emergence of new pathogens and re-emergence of known pathogens in a globally connected world, including the impact of European arrival on Aboriginal and Torres Strait Islander peoples
- scientific and social strategies employed to identify and control the spread of pathogens, including identification of the pathogen and host, modes of transmission and measures to control transmission
- vaccination programs and their role in maintaining herd immunity for a specific disease in a human population
- the development of immunotherapy strategies, including the use of monoclonal antibodies for the treatment of autoimmune diseases and cancer.

KEY KNOWLEDGE

Responding to antigens

The cells of organisms display protein molecules called **antigens**. Antigens represent a recognition code for the organism, identifying the cell as belonging to itself or not. Antigens that belong to the organism's cells are called **self-antigens**. Because **pathogens** (disease-causing organisms) display antigens foreign to the host organisms, they are recognised as **non-self antigens** and are rejected. Pathogens may be described as extracellular or intracellular.

Extracellular pathogens do not invade cells, instead thriving in the extracellular environment. Examples of extracellular pathogens include some bacteria and fungi (Table 4.1.1).

Intracellular pathogens include viruses as well as organisms such as some bacteria and some protozoa that require the internal cell environment to be active.

A **disease** is any condition in an individual that impairs its normal functioning.

- **Infectious diseases** are caused by living organisms or agents of disease.
- Non-infectious diseases are caused by other factors, for example, they may be dietary, inherited or caused by exposure to mutagens.

PATHOGENS

Table 4.1.1 Different kinds of pathogens

Pathogen	Description
prions	• non-cellular • composed of protein • convert cell proteins into prion protein examples: bovine spongiform encephalopathy (BSE, mad cow disease), Creutzfeldt-Jakob disease (CJD)
viruses	• non-cellular • composed of DNA or RNA surrounded by a protein coat • require a host in order to reproduce • structurally much smaller than bacteria • a virion is a single virus particle • a virus that infects a bacterium is called a bacteriophage • unresponsive to antibiotics examples: chickenpox, measles, influenza, HIV/AIDS, Ebola
viroids	• non-cellular • composed of RNA • smallest known infectious particle • plant pathogen
bacteria	• cellular • prokaryotic • feature a cell wall and a circular chromosome • unicellular but may be colonial • can be classified according to structure • respond to antibiotics • can be classified according to response to Gram stain. Gram-positive bacteria respond to certain kinds of antibiotics while Gram-negative bacteria respond to different kinds of antibiotics; this feature of bacteria is important in the development of drugs used to combat infections examples: typhoid, tuberculosis, tetanus, botulism Note: Most bacteria are not pathogenic. cocci: spherical bacteria; bacilli: rod-shaped bacteria; spirochaete: spiral bacteria Types of bacteria
protozoa	• cellular • eukaryotic • unicellular • live in parasitic relationship with host, acquiring nutrients for growth and reproduction from host • may have more than one host organism in its life cycle examples: *Plasmodium falciparum* causes malaria, *Giardia* causes diarrhoea, *Entamoeba histolytica* causes amoebic dysentery Note: Not all protozoa are pathogenic.

ISBN 978 0 6557 0026 5

KEY KNOWLEDGE

Pathogen	Description
parasitic worms	• cellular • eukaryotic • multicellular • live inside host organisms (endoparasites) • live in parasitic relationship with host, acquiring nutrients for growth and reproduction from host • may have more than one host organism in its life cycle examples: roundworm, tapeworm, heartworm
fungi	• cellular • eukaryotic • multicellular or unicellular • heterotrophic: acquire nutrients from the outer layer or dead remains of organisms on which they live examples: tinea, ringworm, thrush, wheat rust
oomycetes	• cellular • eukaryotic • multicellular or unicellular • kingdom Protista • characterised by branching hyphae (haustoria) that penetrate host cells, absorbing nutrients examples: potato blight, *Phytophthora cinnamomi* (an important eucalypt parasite)

Life cycles

Some pathogenic organisms require more than one host to complete their life cycle. The primary host is identified as the host in which the parasite takes on its adult form; the juvenile forms of the parasite occur in the secondary host (Figure 4.1.1).

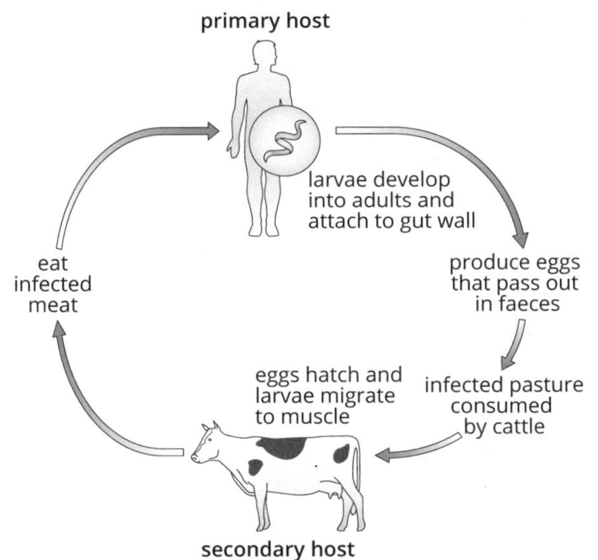

Figure 4.1.1 Life cycle of beef tapeworm

Some parasitic organisms rely on a **vector** for transport from one host to another. For example, although the *Anopheles* mosquito is itself unaffected by the malaria *Plasmodium*, it is the vector that transports the *Plasmodium* from one host to another.

Controlling pathogens

Various antimicrobial agents are available to control the spread of pathogenic organisms. These include:

- antibiotics (taken orally)
- antiseptics (applied to the skin)
- disinfectants (used on surfaces and objects)
- fungicides (kill fungi)
- antifungals (inhibit fungal growth)
- antihelminthics (expel/kill parasitic worms)
- antiprotozoans (kill protozoans).

Although antibiotics are effective against bacteria, they have no effect on viruses.

INNATE IMMUNE RESPONSE

The **innate immune response** (also called the non-specific immune response) includes physical, chemical and microbiota (normal biological flora) barriers that deter the entry of pathogens, as well as a range of strategies for attacking pathogens that have gained entry into the tissues of organisms (Figure 4.1.2 on page 98).

KEY KNOWLEDGE

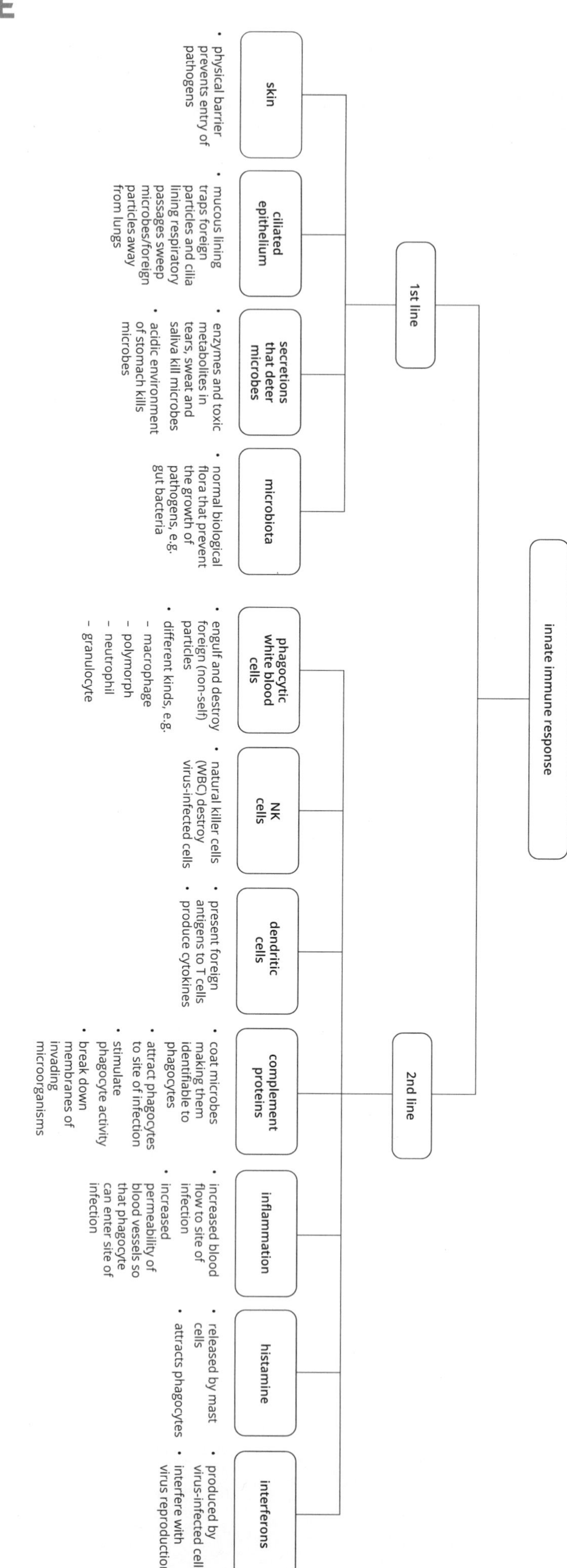

Figure 4.1.2 The innate immune response uses non-specific defence mechanisms.

ISBN 978 0 6557 0026 5

KEY KNOWLEDGE

PLANT DEFENCES AGAINST PATHOGENS

Plants deter pathogens using a range of structural and chemical strategies (Table 4.1.2).

Table 4.1.2 Examples of plant strategies to deter pathogens

Plant feature	Function	Kind of strategy
waxy cuticle coating leaf and stem tissue	physical barrier that keeps pathogens out	structural
lignin	inhibits entry of pathogens; component of woody tissue	structural
production of toxic chemicals such as particular enzymes	breaks down chemicals; promotes apoptosis (programmed cell death)	protein-based chemical
production of toxins such as citronella, tannins	repels insects	chemical
production of defensins	toxic to microbes	chemical

Acquiring immunity

Sometimes pathogens enter the body tissues despite the first-line defences (non-specific barriers) of organisms. For example, some pathogens can gain entry through broken skin, and some through airborne particles breathed into the lungs. When pathogens enter the tissues, second and third-line defence mechanisms come into play.

LYMPHATIC SYSTEM

The mammalian **lymphatic system** is involved in defending the body against disease. It is composed of a transport network of close-ended vessels (similar to the circulatory system's veins) and lymphoid tissues and organs including the bone marrow and thymus gland, spleen, and a series of lymph nodes. The thymus gland and bone marrow are called **primary lymphoid organs and tissues** because they are responsible for the production and maturation of **lymphocytes**. Different kinds of lymphocytes have different roles—some have non-specific roles including circulating through the lymphatic system screening for foreign material (Figure 4.1.3), while others respond in specific ways, such as antibody production. The lymph nodes are rich in white blood cells, including B and T lymphocytes, which are involved in the recognition of foreign antigens.

The **secondary lymphoid organs and tissues** are the lymph nodes, spleen, tonsils, adenoids, appendix and Peyer's patches. Lymphocytes are activated in secondary lymphoid tissues, where they recognise and respond to non-self antigens that are specific to their receptors.

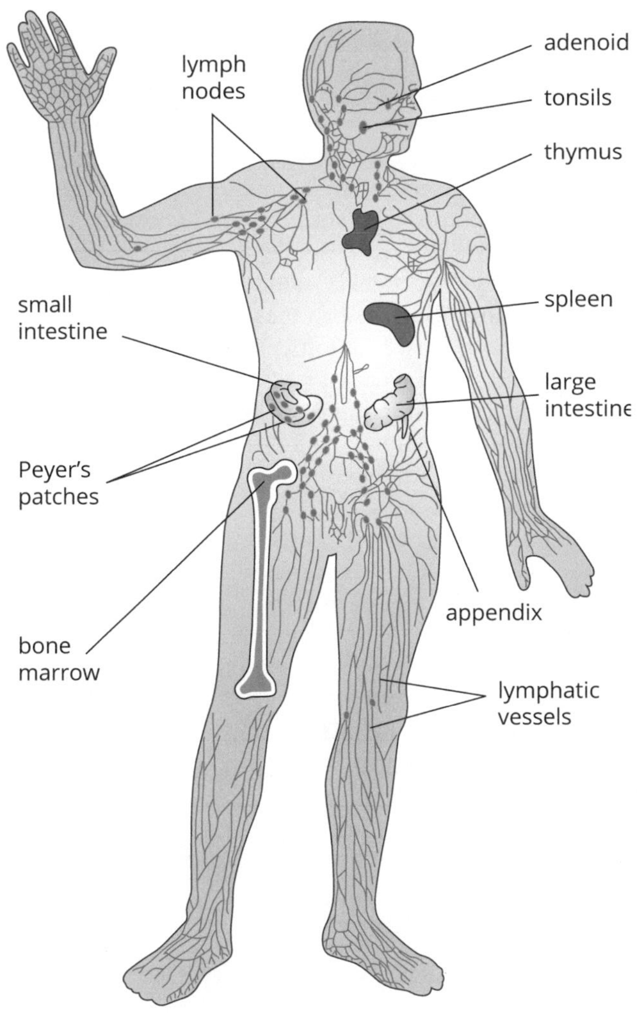

Figure 4.1.3 The human lymphatic system

KEY KNOWLEDGE

ADAPTIVE IMMUNE RESPONSE

The **adaptive immune response** (also called the specific immune response) involves specific defence mechanisms that operate by targeting specific pathogens (Figure 4.1.5). The body's immune system recognises non-self antigens on the surface or plasma membranes of foreign particles, which can include pathogens, allergens and transplant tissue. Specialised lymphocytes called B cells and T cells are involved in recognising foreign antigens. Antigen recognition triggers an immune response that produces antibodies against those specific antigens.

- **Antigen**: protein that stimulates the production of specific antibody; antigens bind with 'matching' antibody.
- **Antibody**: particular kind of protein called **immunoglobulin** produced in response to presence of particular antigen; antibodies bind with antigens (Figure 4.1.4).

The adaptive immune response is characterised by two particular features.

- **Specificity**: antibodies are produced for particular antigens (lock-and-key).
- Memory: **memory B cells** and **memory T cells** are produced after the first encounter with foreign tissue. These memory cells recognise the foreign antigen on subsequent encounters, mounting a defence response more quickly, thereby ensuring no onset of disease symptoms.

Humoral immunity and **cell-mediated immunity** operate together in the adaptive immune response (Figure 4.1.5).

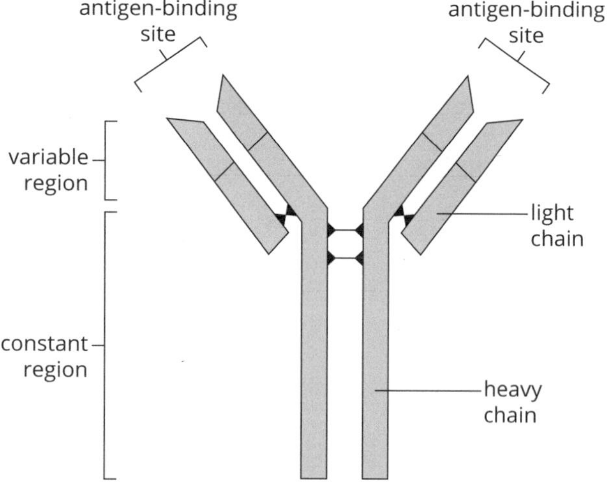

Figure 4.1.4 Antibody

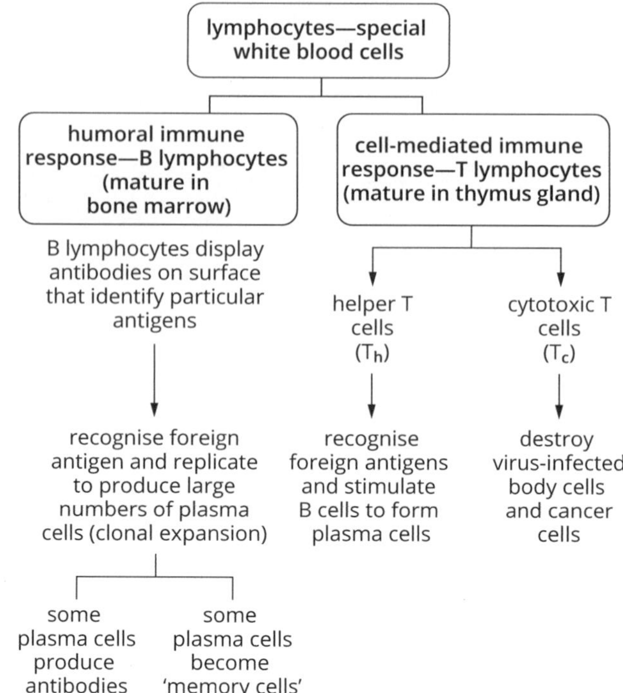

Figure 4.1.5 The adaptive immune response uses specific defence mechanisms.

STRATEGIES FOR ACQUIRING IMMUNITY

Immunity against particular pathogens is achieved in different ways:

- naturally acquired
- artificially acquired
- **active immunity**
- **passive immunity** (Figure 4.1.6).

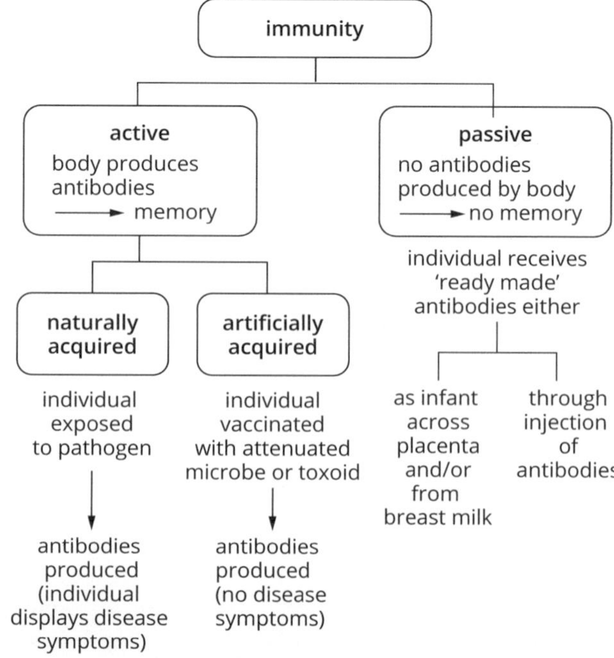

Figure 4.1.6 Immunity against diseases can be acquired actively or passively.

 ISBN 978 0 6557 0026 5

Disease challenges and strategies

The global community and in particular the scientific community is challenged to respond creatively in solving the problem of diseases currently present in the world and to plan responses to manage potential future disease crises.

VACCINATION PROGRAMS

Community **vaccination** programs constitute a significant tool in the control of vaccine-preventable diseases. All children in Australia are routinely vaccinated against serious infectious diseases, including meningococcal disease, pertussis (whooping cough), poliomyelitis, hepatitis and cervical cancer. The incidence of vaccine-preventable diseases has decreased significantly since the introduction of vaccination programs. For example, there have been no reported cases of poliomyelitis in Australia since 1972. The World Health Organization declared smallpox extinct in nature in 1979.

Diseases for which Australians are routinely vaccinated do still occur in the community and when they do they may put vulnerable members of the community at risk of illness and complications including death. Vulnerable members of the community include newborn babies, the elderly, people suffering from a disease that compromises their immune system and people taking immunosuppressant medication.

For immunisation programs to be successful, a sufficient number of people need to be vaccinated. This phenomenon is called **herd immunity**. The more people who are vaccinated, the less chance there is of an infection spreading throughout a population because there will be fewer potential carriers. Herd immunity is essential for the protection of those who cannot be vaccinated or who have suppressed immune systems.

ALLERGIC RESPONSES

An **allergic response** (or allergic reaction) occurs when the body's immune system 'overreacts' to an antigen that has been encountered previously (Figure 4.1.7). The antigen to which the body overreacts is called an **allergen**. Common allergens include pollen and peanuts.

Technological advances in medicine and pharmaceuticals have made treatments available for some allergies. A series of injections of the offending antigen may be used to desensitise the immune system to an allergen.

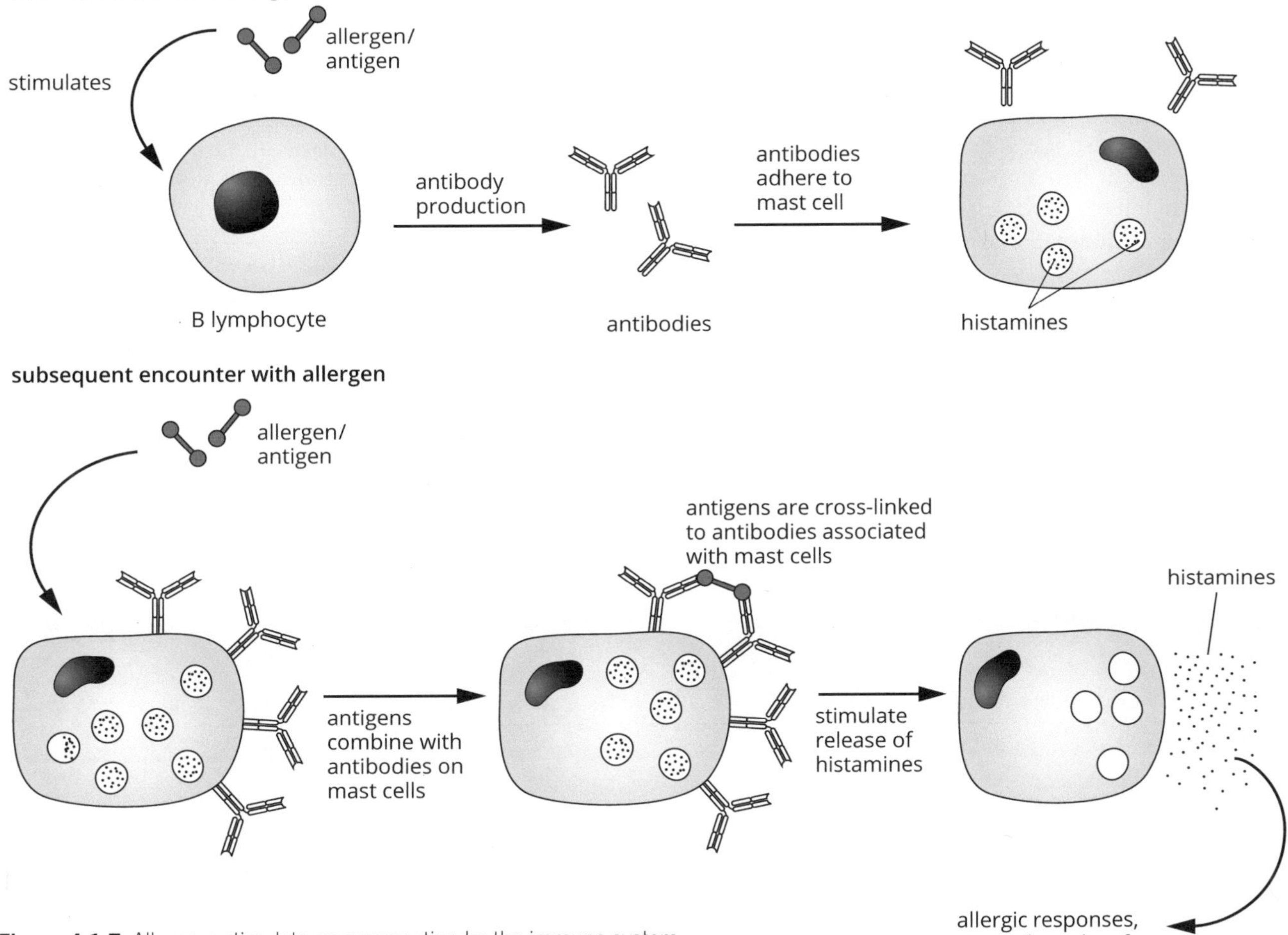

Figure 4.1.7 Allergens stimulate an overreaction by the immune system.

KEY KNOWLEDGE

DEFICIENCIES AND MALFUNCTIONS

Autoimmune diseases occur when the body produces antibodies against its own antigens. That is, the body fails to recognise 'self' cells as 'self', instead recognising them as foreign, and then produces antibodies to destroy them.

Examples of autoimmune diseases are:

- rheumatoid arthritis
- pernicious anaemia
- multiple sclerosis
- type 1 diabetes.

Medical technology continues to make advances into effective treatment for such diseases. For example, type 1 diabetes can be treated in a variety of ways, including insulin therapy and the management of diet. Current research is focusing on an insulin vaccine, transplantation of insulin-producing beta cells in the pancreas, pancreas transplantation and an artificial pancreas.

MHC AND TRANSPLANT REJECTION

The **major histocompatibility complex (MHC)** refers to a group of linked genes that are responsible for the range of antigens present on the surfaces of cells. There are many different types of surface antigens and the combination present on the cells of a particular individual is unique to that individual. This has implications for blood transfusions and transplant surgery. The body's immune system recognises the antigen combination in a similar way to a barcode identifying a supermarket product. If the 'barcode' is identified as non-self, the immune system will produce antibodies against the foreign cells to destroy them. For this reason, it is important to achieve the closest possible match of tissue (antigen combination) between recipient and donor tissue for blood transfusions and transplant surgery.

ABO BLOOD GROUPING AND TRANSFUSION COMPATIBILITY

The ABO blood group system is just one of a number of ways of typing blood. It identifies particular antigens present in the red blood cells. Determining an individual's ABO blood group is achieved by checking for the reaction that occurs between red blood cells and antiserum (antibody preparations).

When red blood cells **agglutinate** in the presence of a particular antiserum, it indicates that there has been a locking together of the surface antigens on the red blood cells and the antibodies in the testing serum. Such a reaction confirms the presence of the particular antigen (Table 4.1.3).

Determining the rhesus status for an individual means checking for the presence of a different type of antigen called the D antigen.

Immunosuppressant drugs are administered to transplant patients to reduce the capacity of the immune system to mount an immune response against the transplant tissue (tissue rejection). Unfortunately, this measure also means that the immune system is less likely to respond to the presence of cancer cells, increasing the chance of some cancers developing.

IMMUNOTHERAPY FOR CANCER

Immunotherapy refers to the manipulation of the body's immune system as a means of responding to diseases such as cancer. The use of **monoclonal antibodies (mAbs)** represents one such example under investigation. 'Monoclonal' refers to clones of a single antibody. In monoclonal antibody therapy, B lymphocytes that produce antibodies specific to the antigen of an identified cancer cell are cloned. When the antibodies are administered to patients, they target only the cancer cells.

EMERGING DISEASES AND INFECTION CONTROL STRATEGIES

With an increasing global population of almost 8 billion people, together with the ease of mobility that modern society enjoys, the management of disease is an ever-increasing challenge.

Epidemics and pandemics

A disease that is limited to local populations is called an **epidemic**. Ebola and Zika virus are examples of epidemics. Outbreaks of the Ebola virus in west African nations and its appearance in some Western countries highlight the problem of the spread of infectious diseases in a globally connected world. While Ebola is highly contagious, it remains largely geographically limited and can be contained with responsible preventative practices in place.

Table 4.1.3 ABO blood groups

Blood sample	Reaction with antibody type			Antigens present	Blood type
	Anti A	Anti B	Anti AB		
1	agglutination	no agglutination	agglutination	A	A
2	no agglutination	agglutination	agglutination	B	B
3	agglutination	agglutination	agglutination	A and B	AB
4	no agglutination	no agglutination	no agglutination	none	O

ISBN 978 0 6557 0026 5

KEY KNOWLEDGE

The Zika virus is a mosquito-borne disease that causes symptoms such as rash and fever, as well as muscle and joint pain. More significantly, it causes brain deformities in unborn babies of infected mothers. Zika virus remains a threat in parts of South America, Africa and Asia. Thanks to strict measures aimed at protecting the community, Australia remains at a relatively low risk for this disease.

A disease that infects populations across a wide geographic area or across continents or the globe is called a **pandemic**. Malaria and COVID-19 are examples of pandemics. The mosquito-borne disease malaria poses a threat to wider populations, with approximately half the world's population at risk. According to the World Health Organization, in 2019 there were around 229 million cases of malaria worldwide, of which 409 000 died.

An outbreak of a new coronavirus, named COVID-19, emerged in China in late 2019. The highly contagious flu-like disease spread quickly, claiming many lives. With infections increasingly recorded around the world, in March 2020 the World Health Organization declared the disease a pandemic.

The crisis led countries around the world, including Australia, to establish strict codes of practice in an effort to curb the spread of the disease. This included measures such as closing international borders to non-citizens and non-residents, implementing physical distancing rules, the wearing of face masks in some situations, work-from-home protocols where possible and remote learning for schools and universities. Community facilities such as cafes, restaurants and gyms were closed for a period.

As well as the Australian federal and state governments implementing varying degrees of these measures, Australian travellers returning from overseas were required to remain in quarantine for a period of 14 days, as was anyone who had come into contact with a person who had been exposed to COVID-19. Some Australian states also closed their borders to domestic travel.

Impact of European arrival on Indigenous Australians

The impact of infectious diseases on individuals and communities is dependent on factors such as age and biological immunity. Populations that have not previously been exposed to a particular pathogen are especially vulnerable. This was the case for Indigenous Australians during the latter part of the 18th century. More than 500 different groups of Indigenous peoples, with an estimated population of 750 000, inhabited the continent prior to British colonisation in 1788. In the 10 years that followed, the impact of introduced diseases was compounded by the dispossession of Indigenous peoples' land, culture and identity. This resulted in conflict over land and resources with the Indigenous population being reduced by as much as 90%. Previously unknown diseases such as measles, influenza and smallpox took a huge toll. While these were relatively mild infectious diseases for Europeans, they had devastating consequences for Australia's Indigenous peoples.

Infection control

Another cause for global concern is the rise of drug-resistant bacteria. When pathogenic bacteria remain unaffected by **antibiotics**, the consequences are serious for human health and the prospect of recovery. Bacteria such as *Staphylococcus aureus* (golden staph) are a classic example of a pathogen displaying multi-drug resistance.

Antibiotics are **antimicrobial drugs** that are effective against bacteria but not viruses. Science has had to apply creative thinking in addressing this significant twenty-first century problem. Hence the emergence of rational drug design—an approach to disease management that bypasses bacteria's ability to develop antibiotic resistance. Examples of rational drug design include chemicals that block the activity of key molecules in the pathogen.

Strategies for preventing the evolution of new pandemic flu strains include regular gene sequencing of human, bird and pig viruses to identify problem variants, the culling of animals that are infected with dangerous new strains and limiting close interaction between humans and animals. Strategies to contain influenza virus include hygiene, wearing face masks to prevent the spread of airborne viral particles, vaccination and **antiviral** medication (Table 4.1.4). Vaccination before each flu season limits the number of infected people. When fewer people and other animals are infected with influenza virus, there is a reduced chance of genetic reassortment giving rise to dangerous new strains.

Table 4.1.4 Antiviral preparations

Drug	Effect
neuraminidase inhibitors	drug molecule binds with viral neuraminidase, an enzyme important in assisting new viral particles from exiting host cells and spreading to infect other cells; viral particles are disabled and the spread of the virus is halted example: Relenza
entry inhibitors	drug molecule binds with part of viral particle that assists entry to cells
reverse transcriptase	drug molecule blocks viral RNA from synthesising DNA
integrase inhibitors	drug molecule blocks viral particles from integrating into host DNA, thereby preventing replication of viral RNA
protease inhibitors	prevents virus from constructing protein coat around new viral RNA

WORKSHEET 19

Knowledge review—responding to pathogens

This activity addresses foundation ideas in biology that you have studied before and on which the key ideas in this area of study are built.

1 The terms set out in the table below are related to organisms' responses to pathogens. It is likely you are familiar with them, especially if you have ever been sick or in need of medication such as antibiotics. Recall your understanding of each of the terms and write down your own definition in the space marked Definition 1. Then use your resources to check your understanding. Rewrite a more formal definition in the space marked Definition 2.

Term		Definition 1	Definition 2
a	disease		
b	pathogen		
c	infectious		
d	contagious		
e	bacterium		
f	virus		

2 When prescribed a course of antibiotics, you will always be advised to take the entire course and not to stop when you start feeling better. Explain the reasoning behind this advice.

ISBN 978 0 6557 0026 5

WORKSHEET 20

Plant and animal defences against invading pathogens

As pathogens rely on invading the tissues and cells of other organisms to acquire resources for survival, the challenge for potential hosts is keeping them out. Plants and animals feature various adaptations to deter pathogens from entry into their tissues and others to combat them if they do manage to enter.

Plants

Plants are susceptible to disease or damage by a range of different kinds of parasitic organisms, including bacteria, fungi, insects, worms such as nematodes, and agents of disease such as viruses. Plants have different defence mechanisms to combat such invasions, including structural and chemical strategies.

1 Complete the following table by describing a potential threat and relevant defence strategy for each example, and identify the strategy as structural or chemical in each case.

Example		Threat	Defence strategy	Structural or chemical mechanism
a	cell wall, waxy cuticle, bark waxy cuticle			
b	caterpillar infestation on a *Eucalyptus* tree			
c	wheat with smut fungi			

	Example	Threat	Defence strategy	Structural or chemical mechanism
d	spinach with leaf miner larvae			

Animals

While the defence mechanisms of different kinds of animals depends on the organism in question, one first-line defence feature common to all animals is their skin. Intact skin forms a physical barrier that deters the entry of pathogens in the first place. Other first-line defence mechanisms include chemical barriers such as mucus, antibacterial solutions, pH and the action of cilia.

2 Complete the boxes below by identifying the different kinds of defence mechanism that apply, describing the role of each. A couple have been started for you.

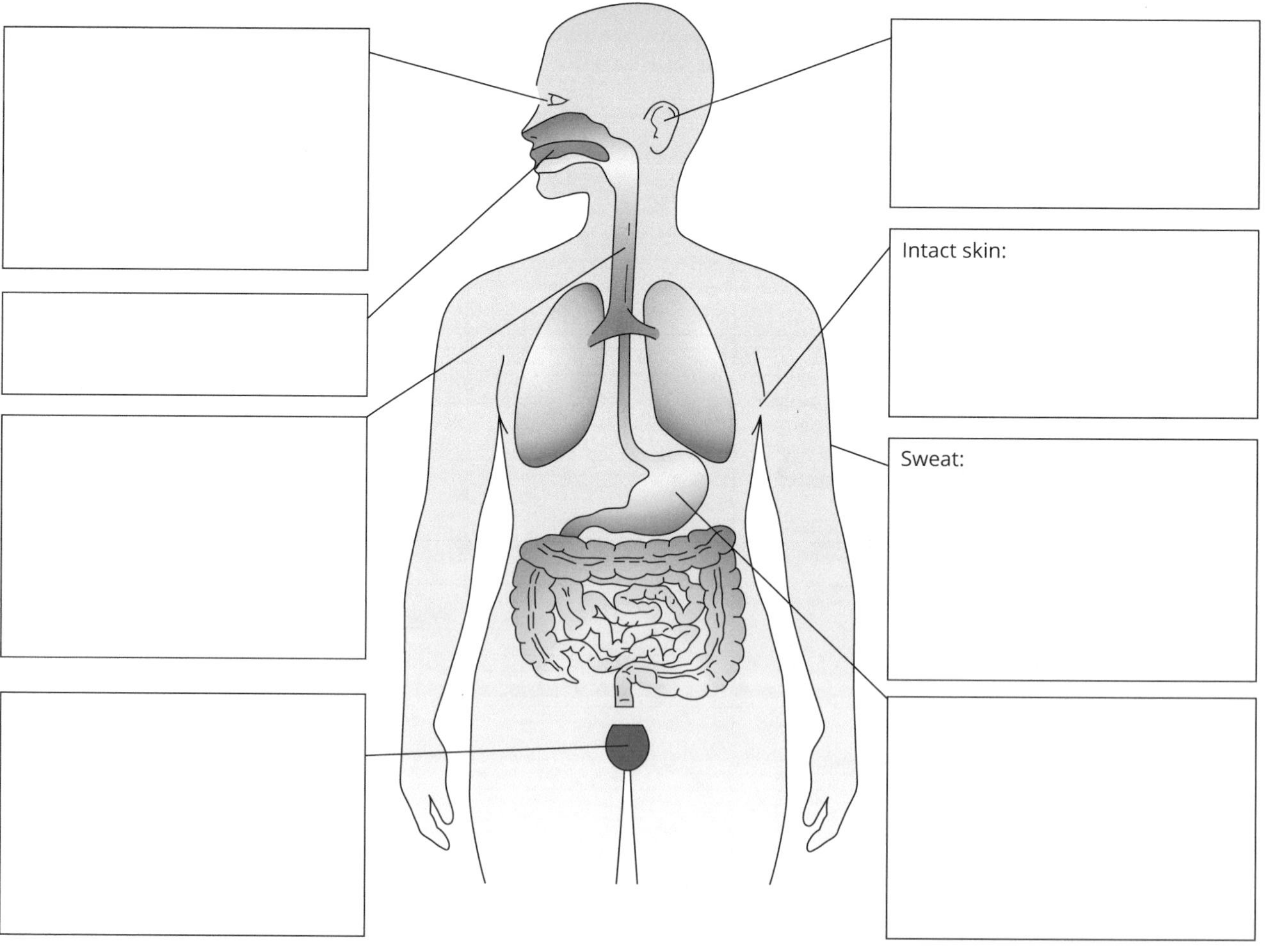

 ISBN 978 0 6557 0026 5

WORKSHEET 21

Second- and third-line defence mechanisms in humans

1 Make a selection from the list below to fill in the missing words. You will then have a completed summary of the way in which the body defends itself against invading pathogens.

memory	mast cells	interferons	lymphatic system	transplant
inflammation	antiseptic	phagocytic	specific	physical
specific adaptive	T lymphocytes	complement	plasma	antibodies
ciliated	cytokines	NK cells	antigens	humoral

a Organisms have evolved different means of defending themselves against pathogens. These can be non-specific or ________________. Non-specific defence mechanisms include:

- external surfaces, like skin, that provide a ________________ barrier to entry of pathogens
- ________________ present in sweat, tears and saliva that destroy bacteria
- ________________ membranes, e.g. lining airways, that sweep foreign material away
- ________________ cells target, ingest and destroy foreign particles
- proteins that make up the ________________ system—they trigger reactions such as encouraging phagocytes, destroying bacteria and contributing to ________________
- specialised molecules called ________________ interfere with virus replication.

b Vertebrates have evolved a complex system of targeting and destroying particular pathogens in immune responses. This is called ________________ ________________ immunity. The two key features of this system are specificity and ________________.

c Immune responses rely on two lines of defence that operate together. These are ________________ immunity that involves B lymphocytes or B cells, and cell-mediated immunity that involves ________________ or T cells.

d In specific adaptive immunity, B lymphocytes recognise foreign ________________ present on invading bacteria, viruses and toxins. They produce ________________ specific to the antigen.

e With the assistance of helper T cells, B cells form two kinds of daughter cells: memory cells that quickly identify the pathogen should a subsequent infection occur; and ________________ cells that produce antibodies.

f There are two kinds of T lymphocytes, each with a particular function. Helper T cells produce cytokines that control the function of other B cells and T-cytotoxic cells. T-cytotoxic cells act against virus-infected body cells and cancer cells. They are also involved in the rejection response against ________________ tissue.

2 You should have four terms left over. Write out each of these terms together with their definitions.

____

____

____

____

WORKSHEET 22

Specific adaptive immunity

When invading pathogens successfully enter the body and its tissues, the immune system responds by mounting an attack that specifically targets the pathogen. Antigens on the surface of the invader are recognised as non-self. This triggers a series of events and processes aimed at destroying the invader.

Complete the flow chart below to build a picture of how humoral and cell-mediated responses operate to rid the body of an invading pathogen and develop immunity against it in the event of subsequent exposure.

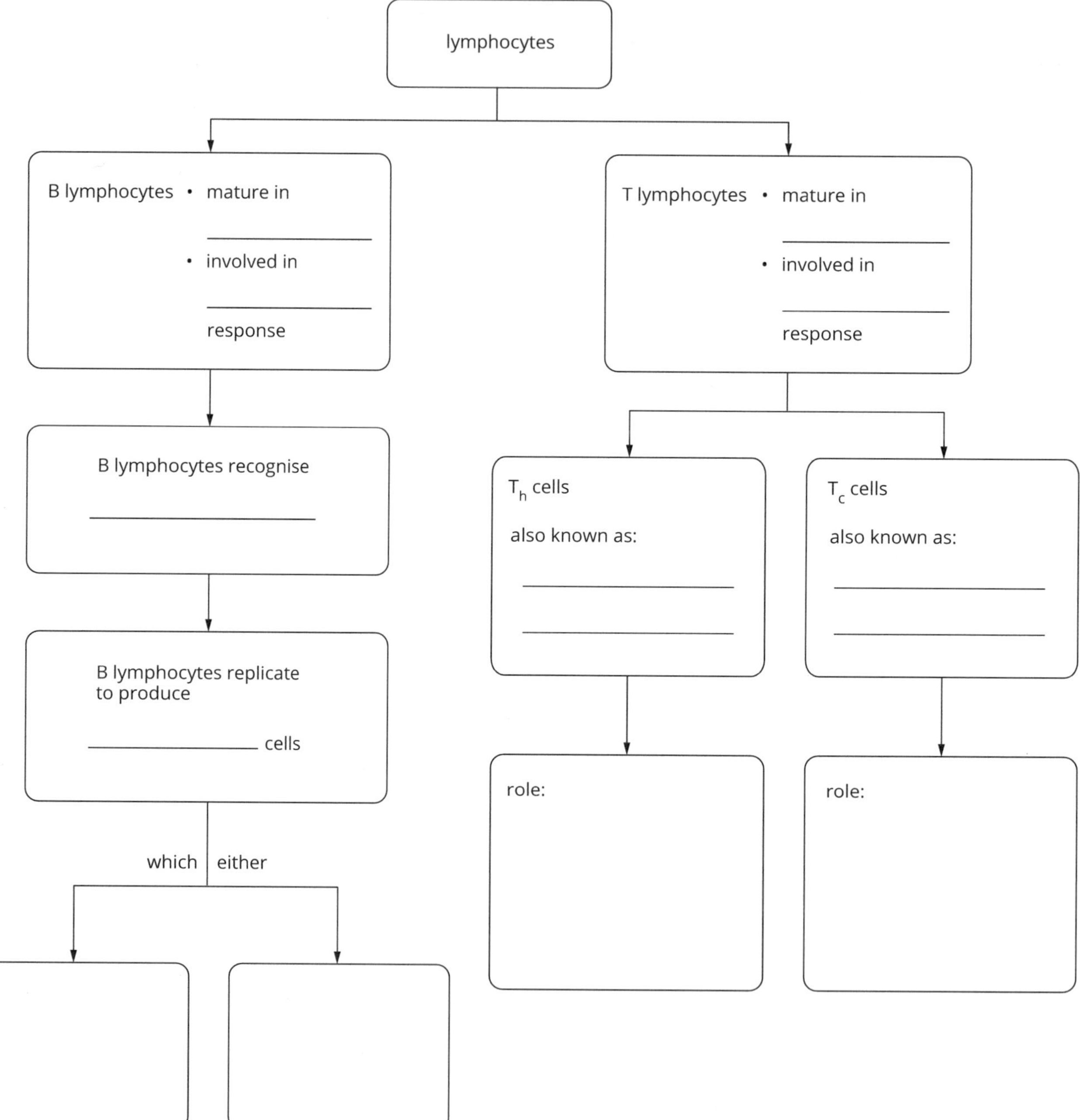

 ISBN 978 0 6557 0026 5

WORKSHEET 23

An overview of immunity

1 Read the definitions listed in the boxes on the right side of the page. Choose the correct term from the list below to match each definition. Write each term in the box corresponding to its definition.

antibody	passive immunity	antihistamine	immunoglobulin	histamine
'non-self'	antigens	adaptive immunity	vaccine	artificially acquired active immunity
'self'	autoimmune disease	allergy	toxoid	allergen	naturally acquired active immunity

Term	Definition
	immunity against a pathogen that develops in response to infection by the pathogen
	substance released by mast cells that promotes inflammation, dilation of blood vessels and leakage of phagocytes into damaged tissue
	describes recognition proteins on the cell surface accepted by the body as belonging to itself
	molecules that combine with antibodies and stimulate immune responses
	body's ability to defend itself against specific pathogens
	disease in which the body fails to recognise 'self' antigens and mounts an immune response against its own tissues
	overreaction by the immune system to a previously encountered antigen
	specific protein produced by the body in response to a foreign antigen and that combines with the foreign antigen
	immunity conferred by the administration of antibodies
	preparation of dead or attenuated pathogenic particles that provoke an immune response

2 You will see that there are six terms left over. Write a definition for each of these terms.

WORKSHEET 24

Case study

Vaccination programs and polio

You may not have heard of poliomyelitis, but you have almost certainly been vaccinated against it. Polio is a viral infection that affects neuron function, which in turn affects coordination. While most infected individuals do not present with symptoms, those who do report headache, fever, nausea and vomiting. Only 2% of patients suffer from more severe symptoms of muscle pain and stiffness, typically in the back and neck. Less than 1% of infected individuals show complications, which include severe muscle weakness, paralysis and death. Patients recovering from severe infection may have withered limbs and require mobility assistance.

Polio is extremely contagious and easily transmitted by close contact or airborne droplets from an infected person. The last reported case of polio in Australia occurred in 1972. Australia has high immunisation rates for poliomyelitis and was declared polio-free in the year 2000. Children in Australia are routinely immunised against polio at two, four and six months, with a booster at about four years of age.

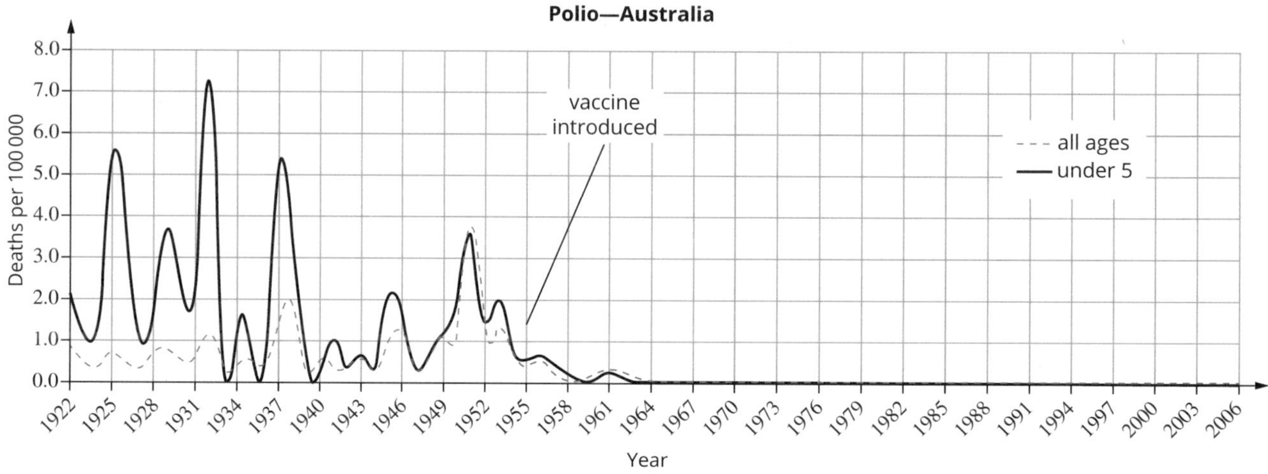

Figure 4.1.8 The impact of the introduction of the polio vaccine in Australia

1 **a** Referring to Figure 4.1.8, in what year did the immunisation program against poliomyelitis begin in Australia?

b Describe the mortality statistics for polio in the years before and after this date.

2 Use internet resources to find out the treatment for polio.

3 Justify the need for a nationwide immunisation program for polio given the small percentage of people affected by it.

4 If Australia has been declared polio-free, suggest why the immunisation program for this disease continues.

ISBN 978 0 6557 0026 5

5 Polio is a notifiable disease. This means that suspected cases of the disease must be reported to health authorities.

a Suggest reasons for this regulation.

b Suggest the likely management for a person being tested for polio.

WORKSHEET 25

Modelling

Allergic responses

Hay fever is an example of an allergic response that occurs when the body overreacts to an otherwise harmless environmental factor that it has encountered previously.

1 Use the symbols shown and your knowledge of allergies and allergens to complete the sequence of steps that occurs in an allergic response. You will need to add to the picture steps, label the different components and provide explanatory notes to complete the process.

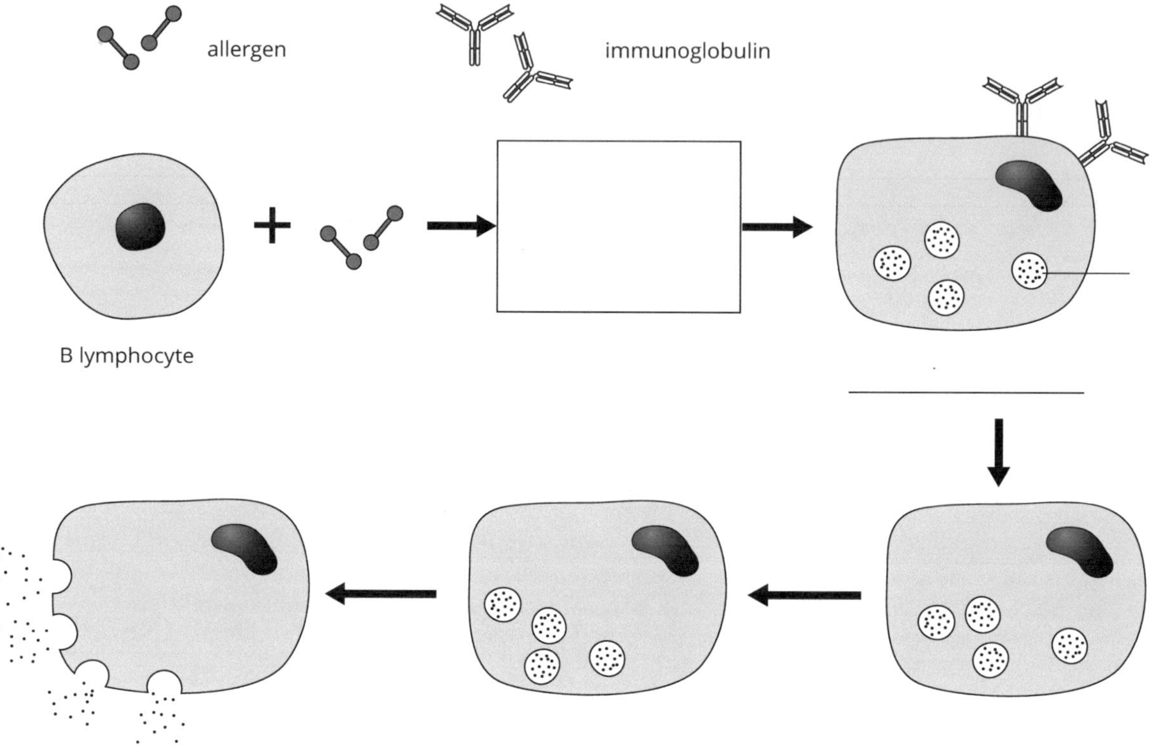

2 Write down another term that is interchangeable with the term 'immunoglobulin'.

3 Where in the body are mast cells located?

4 Describe how the release of histamines contributes to the symptoms of an allergic response such as hay fever.

5 Explain how injecting allergen particles may desensitise an individual to a particular allergen.

 ISBN 978 0 6557 0026 5

WORKSHEET 26

Modelling

Antibodies and immunity

Antibodies are specific proteins called immunoglobulins produced by the body's immune system in response to the presence of particular antigens. When antigens recognised as foreign or 'non-self' are detected, the immune system mounts a response, producing antibodies with complementary binding sites. Antibodies bind with the antigens to disable invading pathogens.

1 The figure below represents a typical antibody.

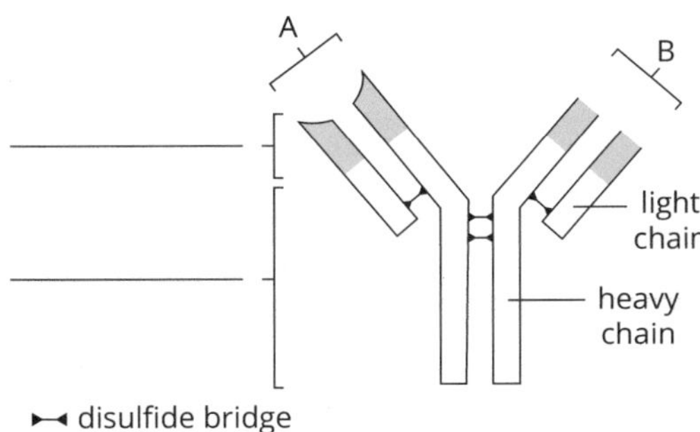

a Correctly label the variable region and the constant region of the antibody shown.

b Sites A and B indicate the antigen-binding site of the antibody. B is incomplete. Complete the diagram to show the correct shape of the antigen-binding site at B for this antibody.

c Explain your choice of shape in part **b**.

2 Vaccination is one way of achieving artificially acquired active immunity. Vaccines contain pathogenic antigens that stimulate the immune system to mount the same antibody response that would occur during exposure to the actual disease.

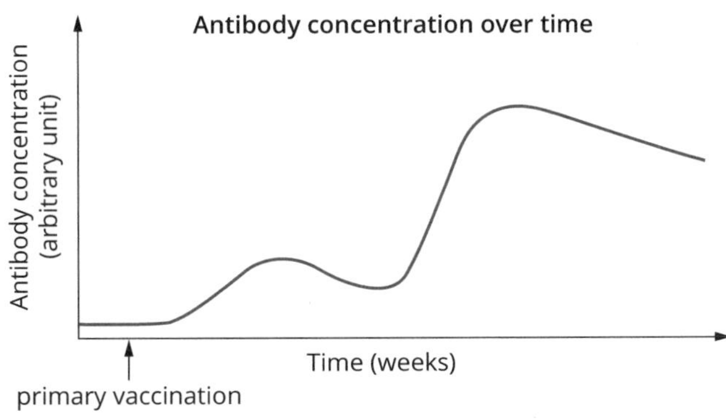

a Insert an arrow along the *x*-axis of the graph above to show where a secondary vaccination would have been given. Explain your choice.

b Describe two features of the secondary immune response compared with the primary response.

c Vaccines involve the production of 'attenuated pathogens'. Explain what is meant by this term.

d Explain why the antibody concentration shown on the graph does not begin at zero prior to the primary vaccination.

e Outline an advantage of using live vaccines rather than administering specific antibodies.

f If exposure to actual pathogens is enough to stimulate antibody production in individuals, why do we bother with vaccination programs?

 ISBN 978 0 6557 0026 5

WORKSHEET 27

Reflection—How do organisms respond to pathogens?

The following table lists the key knowledge covered in this area of study.

1 Reflect on how well you understand the concepts listed. Rate your learning by shading the circle that corresponds to your current level of understanding for each one.

Key knowledge	Not confident ◄			► Very confident
Physical, chemical and microbiota barriers that prevent infection in plants and animals	○	○	○	○
Innate immune responses to pathogens	○	○	○	○
The lymphatic system	○	○	○	○
Adaptive immune responses to pathogens	○	○	○	○
Naturally and artificially acquired immunity	○	○	○	○
The emergence of new pathogens and re-emergence of known pathogens	○	○	○	○
Scientific and social strategies to identify and control the spread of pathogens	○	○	○	○
Vaccination programs and herd immunity	○	○	○	○
Immunotherapy strategies	○	○	○	○

2 Consider the points you have shaded from Not confident to Very confident. List specific ideas you can identify that were challenging.

3 Write down two different strategies that you will apply to help further your understanding of these ideas.

PRACTICAL ACTIVITY 9

Controlled experiment

Investigating the effect of antibiotics on bacterial growth

Suggested duration: 30 minutes on first lesson; 60 minutes on second lesson

INTRODUCTION

Mastrings or Multodiscs are commercially produced discs impregnated with a variety of different antibiotic substances. Antibiotics are chemicals that deter the growth of bacteria. Different kinds of bacteria are affected by different kinds of antibiotics.
Each satellite on the Mastring contains a different antibiotic that is identified by either a colour or a letter code or both.

MATERIALS

- disposable gloves
- 3 sterile, prepared nutrient agar Petri dishes
- broth cultures of *Escherichia coli* (*E. coli*) and *Staphylococcus albus* (*S. albus*)
- 2 sterile swabs in container
- marker pen
- fine forceps
- clear adhesive tape or Parafilm M laboratory tape
- access to incubator set at 35°C
- 2 sterile Mastrings
- disposal bags
- paper towel
- disinfectant
- access to soap and water for washing hands

AIM

- To investigate the effect of antibiotics on bacterial growth.

METHOD

Caution: Do not open the Petri dishes once sealed. Follow guidelines for safe disposal.

1 ▪ Put on disposable gloves. Collect three sterile nutrient agar Petri dishes. Keep the lids in place and use a marker pen to label the three dishes A, B—*E. coli*, and C—*S. albus*. Use initials or some other code to identify your Petri dishes from others in the class. It is a good idea to label the underside of the dish only, and at the perimeter rather than the middle.

2 ▪ Collect a container of each of the two bacteria and two sterile swabs. Your teacher will demonstrate how to prepare a 'lawn culture' of the bacteria. Use the swabs to prepare bacterial cultures of *E. coli* and *S. albus* in Petri dishes B and C respectively. Immediately replace the lids. Place swabs in a disposal bag.

3 ▪ Collect two Mastrings. Use fine forceps to place one Mastring in the centre of Petri dish B, as shown in Figure 4.1.9. Replace the lid. Repeat this procedure for Petri dish C.

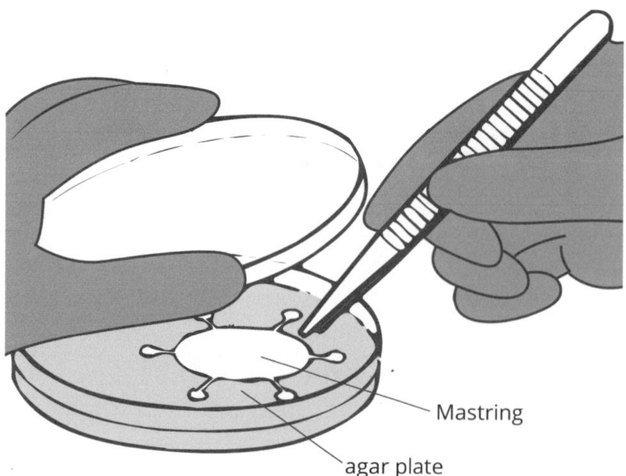

Figure 4.1.9 Mastring on agar plate

4 ▪ Seal each of the Petri dishes with four short pieces of clear adhesive tape, placed at 12 o'clock, 3 o'clock, 6 o'clock and 9 o'clock. Alternatively, seal the Petri dishes with a strip of Parafilm M laboratory tape.

5 ▪ Set the Petri dishes in place to be incubated at 35°C for 24–36 hours or at room temperature for three days. This will depend on your class timetable and teacher discretion.

6 ▪ When all equipment is put away, wipe down your bench using paper towel and disinfectant. Then wash your hands thoroughly.

ISBN 978 0 6557 0026 5

PRACTICAL ACTIVITY 9

1 Explain the purpose of Petri dish A.

2 Suggest a hypothesis being tested by this activity.

Next lesson

7 ▪ Collect your three Petri dishes for observation.
DO NOT OPEN THE DISHES.

3 Use the circles below to prepare detailed drawings of what you see in each Petri dish. Include appropriate labelling as well as a description of your observations.

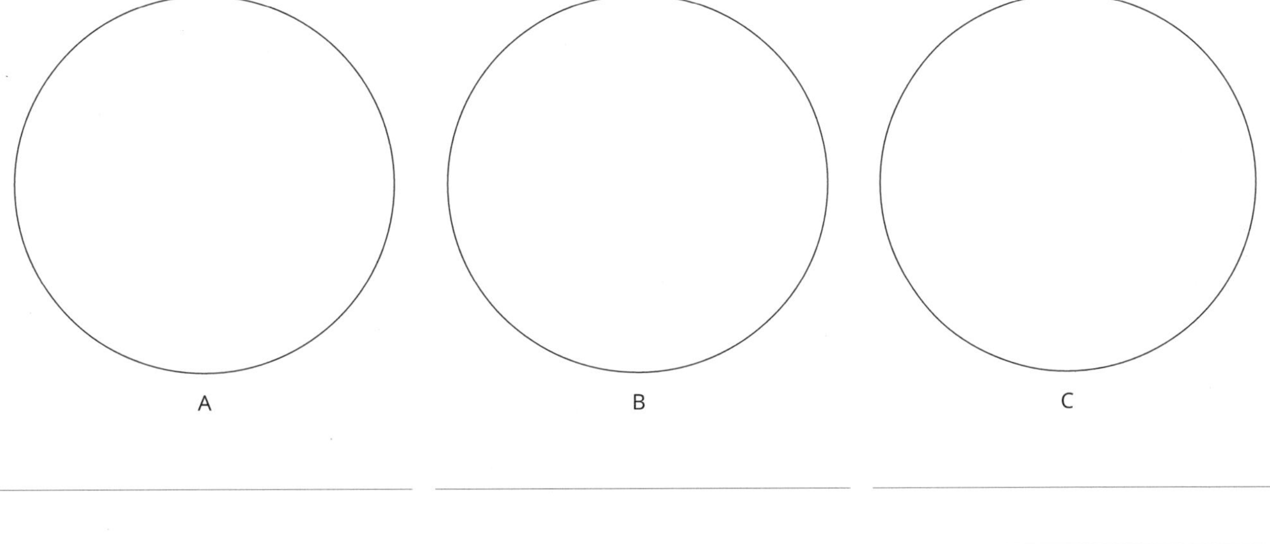

4 Examine Petri dishes B and C that include the Mastring. Use your observations to complete the following table.

Record of results			
Disc code	**Antibiotic on disc**	**Observations of bacterial growth around disc (diameter of clear zone, mm)**	
		E. coli	***S. albus***

ISBN 978 0 6557 0026 5

5 Why is it important to ensure the Petri dishes remain sealed?

6 Discuss any limitations related to this activity. Outline how each limitation might affect the investigation and make suggestions about how to improve this for subsequent investigations.

7 Compare the sensitivity of the bacterial species to the range of antibiotics used.

8 ▪ Place all Petri dishes in the disposal bag provided. Wash your hands thoroughly.

8 In this activity you have used both antibiotics and disinfectant.

a Outline the similarities and differences between antibiotics and disinfectants and their uses.

b Antiseptics are a different kind of antibacterial preparation. Describe the purposes for which antiseptics are used.

CONCLUSION

9 Summarise your findings in this activity. Include whether or not your hypothesis was supported, drawing on the experimental evidence to support your statement.

 ISBN 978 0 6557 0026 5

PRACTICAL ACTIVITY 10

Classification and identification • Modelling

Vaccination programs and herd immunity

Suggested duration: Part A—30 minutes; Part B—30 minutes; Part C—30 minutes

INTRODUCTION

Immunisation programs in Australia are routine measures set in place to protect individuals and the community at large from infection by life-threatening pathogens. Immunisations involve the administration of attenuated vaccines that contain antigens from the pathogens in question. The introduction of foreign antigens stimulates the body's immune system to produce antibodies specific to each kind of pathogen (Figure 4.1.10). When the immune system produces its own antibodies, it retains a 'memory' of the pathogen, which launches more rapid antibody production in the event of subsequent exposures.

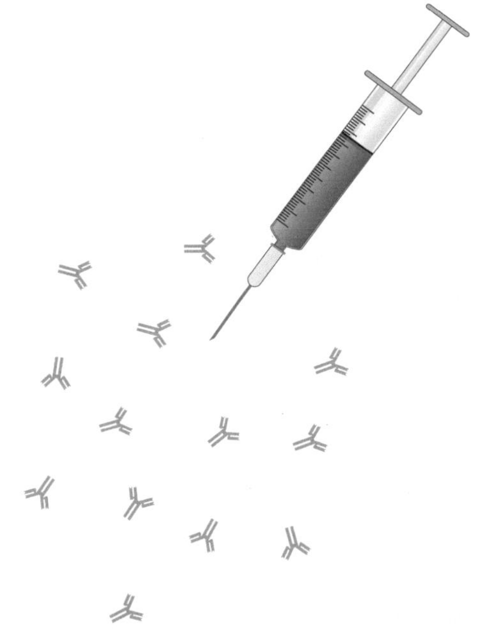

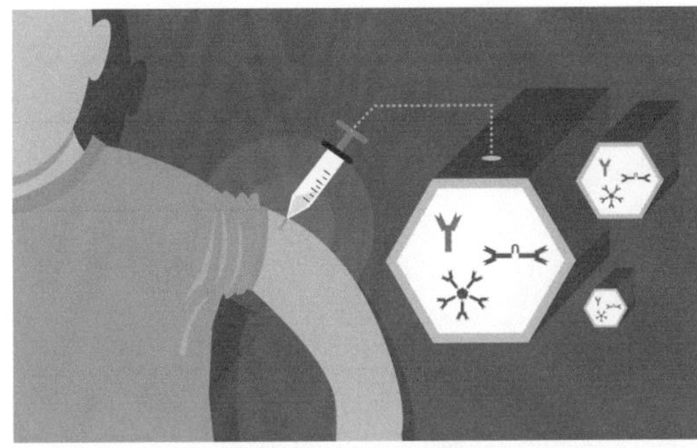

Figure 4.1.10 Vaccines contain antigens that stimulate the body's immune system to produce antibodies specific to a pathogen.

An infant's first vaccination is typically administered at two months of age and followed with repeat vaccinations at four and six months. Further vaccinations occur at 12 months, 18 months and continue at prescribed intervals throughout childhood, with boosters recommended periodically after this. Australians are routinely vaccinated to provide protection against the following serious diseases:

- hepatitis B
- diphtheria
- tetanus
- pertussis (whooping cough)
- haemophilis influenza b
- pneumococcal conjugate (streptococcal pneumonia)
- poliomyelitis
- rotavirus
- meningococcal disease
- measles
- mumps
- rubella
- varicella (chickenpox)
- human papillomavirus
- influenza

AIM

- To consider the different ways of acquiring immunity against pathogens.
- To examine statistics related to deaths from vaccine-preventable diseases.
- To model the transmission of an infectious disease in a community.

PART A • PATHWAYS TO ACQUIRING IMMUNITY

There are different ways individuals can acquire immunity to pathogens. These may involve exposure to the pathogen, vaccination or acquiring ready-made antibodies. Some are long-lasting while others provide relatively short-lived immunity.

1 The following table summarises various aspects of the different ways in which individuals may acquire immunity against diseases. Complete the table by adding notes according to the following instructions.

- **a** Define active and passive immunity.
- **b** Describe how the active and passive immunity listed in part **a** can be acquired naturally or artificially.
- **c** Provide an example of how each kind of immunity described in part **b** can be achieved.
- **d** Describe an advantage for each kind of immunity described in part **b**.
- **e** Describe a disadvantage for each kind of immunity described in part **b**.

Ways of acquiring immunity				
Acquired immunity				
a	Active immunity:		Passive immunity:	
b	Naturally acquired:	Artificially acquired:	Naturally acquired:	Artificially acquired:
c	Example:	Example:	Example:	Example:
d	Advantage:		Advantage:	
e	Disadvantage:		Disadvantage:	

2 Infants under two months old are typically unimmunised against any of the serious diseases listed in the introduction. Explain why newborn babies are still likely to have some antibody protection against such diseases.

 ISBN 978 0 6557 0026 5

3 Explain what is meant by the term 'attenuated vaccine'.

4 Explain why separate vaccines need to be prepared for each different kind of communicable disease.

PART B • VACCINATION PROGRAMS

Table 4.1.5 summarises immunisation data including deaths resulting from some serious communicable diseases in Australia between 1926 and 2000.

Table 4.1.5 Deaths from vaccine-preventable diseases

Period	Diphtheria	Pertussis	Tetanus	Poliomyelitis	Measles	Population estimate
1926–1935	4073	2808	879	430	1102	6600000
1936–1945	2791	1693	655	618	822	7200000
1946–1955	624	429	625	1013	495	8600000
1956–1965	44	58	280	123	210	11000000
1966–1975	11	22	82	2	146	13750000
1976–1985	2	14	31	2	62	14900 000
1986–1995	2	9	21	0	32	17300 000
1996–2000	0	9	5	0	0	18734 000

indicates decade in which community vaccination started for the disease

5 Identify the decade in which the death rates were highest for:

a diphtheria

b pertussis

c poliomyelitis

6 Contrast the population growth between 1926 and 2000 against the pattern of disease incidence in the same time.

7 Identify the event that was instrumental in reversing the death rate in Australia resulting from these diseases.

8 Suggest why the incidence of these diseases did not drop to zero in the period after community immunisation programs commenced.

9 The fatality rate varies for these different serious diseases. Suggest why some infected individuals die while others survive.

PART C • HERD IMMUNITY

Whooping cough is a classic example of an illness that illustrates the significance of herd immunity and the grave consequences when herd immunity is not in place. It is a contagious and serious respiratory illness caused by the bacterium *Bordetella pertussis*. In its early stages patients present with symptoms similar to the common cold, including a runny nose, fever, fatigue and a cough. It is characterised by severe bouts of coughing with a distinct 'whooping' sound as patients draw breath. The condition is most dangerous for infants, with complications including inflammation of the brain, haemorrhage, apnoea (breathing is halted for periods of time), convulsions, brain damage and death. According to the Royal College of Pathologists of Australasia statistics, 2808 people in Australia died as a result of complications of whooping cough in the decade from 1926–1935, with a further 1693 deaths between 1936 and 1945.

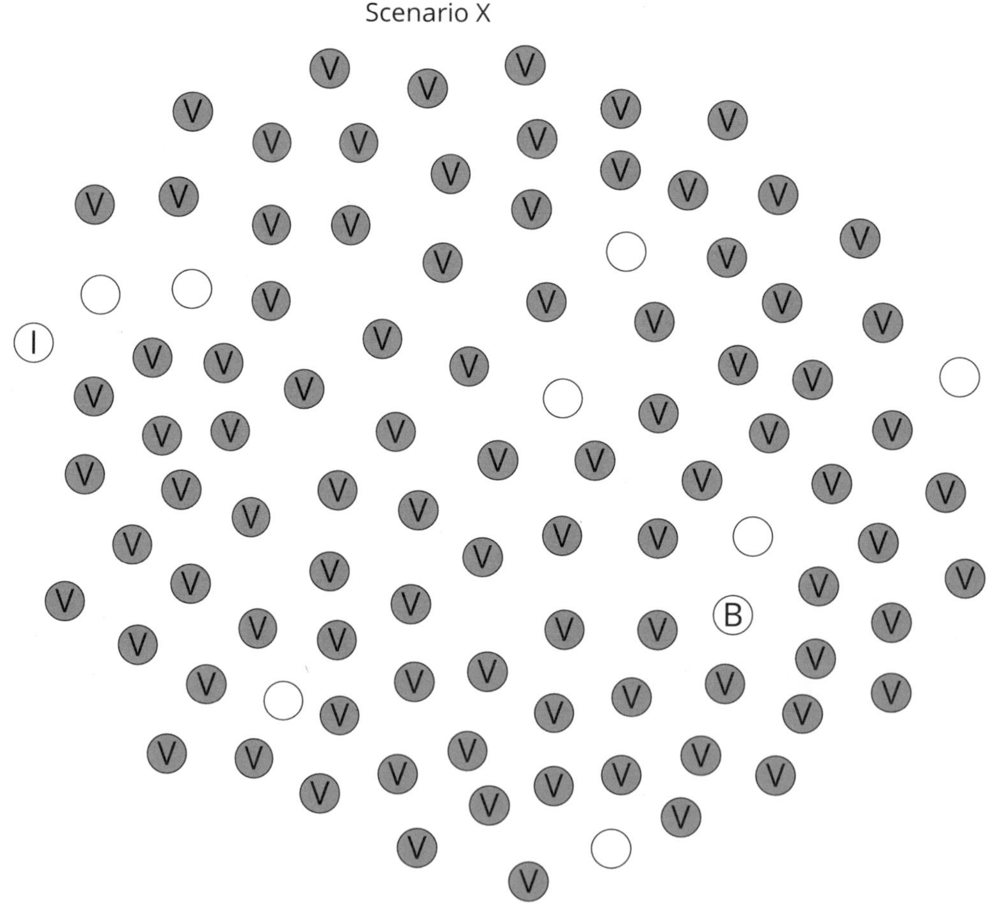

ISBN 978 0 6557 0026 5

PRACTICAL ACTIVITY 10

By the end of 1945 community vaccination programs were in place—this was followed by a dramatic decrease in whooping cough cases and deaths. In recent years there has been on average less than one death per year in Australia from whooping cough, however the sad statistic remains that one in 200 infants that contracts the disease dies. The most effective means of reducing the risk of infection to whooping cough is community vaccination programs. As the immune system of infants is still maturing, they do not receive their first vaccination against whooping cough until two months of age.

Community vaccination programs offer an important strategy in protecting its members from infection, particularly vulnerable members such as unimmunised infants. Herd immunity depends on most members of the community being vaccinated. The more contagious a disease is, the higher the percentage of community vaccination required to protect those not vaccinated. Whooping cough is very contagious. The threshold of community vaccination to protect the herd is between 90–95%.

10 Scenarios X and Y represent simplified models of the transmission of whooping cough infection in two similar communities. In each case, use the symbols ⟶ and —< as shown in the legend to track both the pathway of infection and non-infection for this disease. Use a red pen to show the infection pathway ⟶ and blue or black for the non-infection pathway —<. Start with individual Ⓘ. Note that only infected individuals are able to pass on the disease. Vaccinated individuals are themselves protected and in turn are unable to infect others, increasing protection in the community.

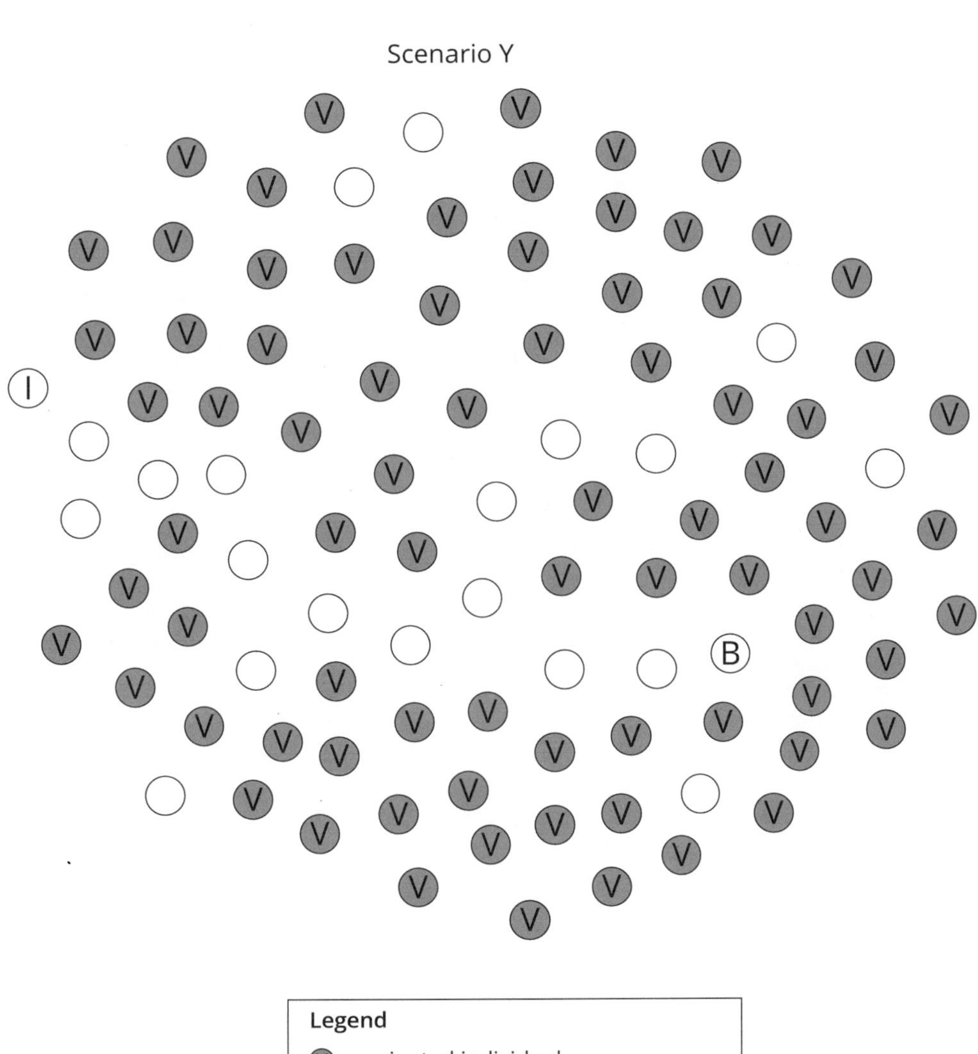

Legend

- Ⓥ vaccinated individual
- ◯ unvaccinated individual
- Ⓘ infected individual—whooping cough
- Ⓑ baby—under 2 months old
- ⟶ pathway of infection
- —< pathway of non-infection

PRACTICAL ACTIVITY 10

11 Explain what is meant by 'herd immunity'.

12 Which scenario, X or Y, represents herd immunity? Justify your choice.

13 Individual B is a newborn baby. Compare the whooping cough risk to baby B in both scenarios. Explain fully.

14 **a** Despite Australia's rigorous vaccination programs, infants are periodically exposed to whooping cough. Some of these babies die. Why are infants, especially those less than two months old, susceptible to diseases such as whooping cough?

b Describe possible sources of infection for these infants.

15 How does herd immunity offer protection against whooping cough to infants?

16 There are some rare examples of individuals in the community who are advisedly exempt from vaccination. Suggest the circumstances that might lead to such exemption.

ISBN 978 0 6557 0026 5

CONCLUSIONS

17 Summarise the importance of active immunity in protecting individuals from serious communicable diseases.

18 Outline the importance of herd immunity in protecting vulnerable members of the community against communicable diseases.

19 Comment on the impact and importance of immunisation programs on the incidence of serious communicable diseases in Australia.

PRACTICAL ACTIVITY 11

Modelling • Simulation

Cutting-edge strategies to combat disease

Suggested duration: 60 minutes

INTRODUCTION

The use of chemical agents against pathogens revolutionised healthcare in the 20th century. Prior to the 1930s, even seemingly minor infections could have devastating consequences for individuals, sometimes resulting in death. Then in 1932 the first sulfur-based antibiotics were discovered. By 1936 antibiotics were already widely used in medicine, saving millions of lives. Antibiotics are medicines used to treat infections caused by bacteria. Bacteria can affect the body in different ways, including inducing fever and producing toxins that affect cell functioning. Bacteria can reproduce at a rapid rate—in optimal conditions *E. coli* can divide every 20 minutes. Antibiotics may work by:

- breaking down the bacterial cell wall, which kills the bacteria (bactericidal). Penicillin is a classic example of an antibiotic that works in this way
- blocking nutrients from getting to the bacteria, thereby halting essential cellular processes and stopping the bacteria from multiplying (bacteriostatic).

Bacteria can be grouped into two broad categories based on the chemical structure of their cell walls. These two groups can be identified by the way they respond to Gram stain. Gram-positive bacteria stain purple as their cell walls absorb the dye. Gram-negative bacteria stain pink due to a different and more complex cell wall structure. Identifying the bacteria by their cell wall is important in determining how infections are managed.

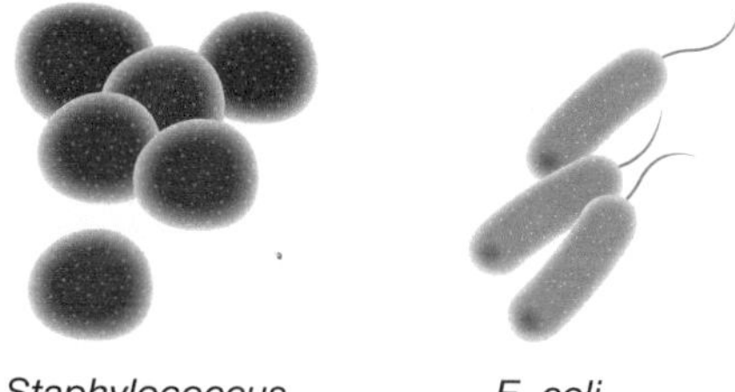

Figure 4.1.11 Two different kinds of bacteria (*Staphylococcus* = coccus; *E. coli* = bacillus)

Some antibiotics, such as penicillin, are effective against Gram-positive bacteria but are not absorbed by the cell wall of Gram-negative bacteria. Gram-negative bacteria respond to different kinds of antibiotics. Tissue swabs taken from infected individuals are used to identify whether pathogenic bacteria are Gram-positive or Gram-negative, which in turn determines which antibiotics should be prescribed.

In recent years the overuse of antibiotics has seen the development of drug-resistant strains of pathogenic bacteria. Examples include *E. coli*, which can cause urinary tract infections and the sexually transmitted disease gonorrhoea. Drug resistance in pathogenic bacteria poses a significant problem for modern medicine and challenges researchers to develop new and creative solutions to the management of infectious diseases.

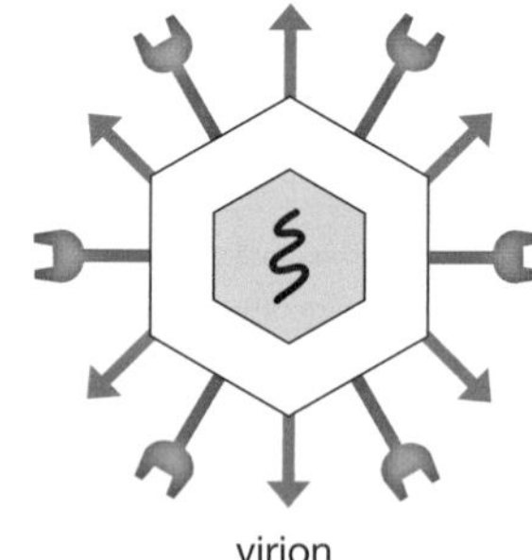

Figure 4.1.12 A single virus particle illustrating nucleic acid core surrounded by protein coat and entry/exit molecules

In the 1980s, antiviral drugs were being developed for clinical use to deal with a different type of pathogen—viruses (Figure 4.1.12). Viruses are non-cellular agents of disease and are unable to reproduce independently; to reproduce they must invade a host cell, using the cell's DNA to make copies of their own viral DNA or RNA. Once multiple copies of the viral machinery are produced, the virus particles leave the cell and spread to other cells to repeat the process.

A range of antiviral drugs is available today. They may act against viruses by:

- blocking the binding site of the virus to cells, thereby preventing virus particles from entering cells
- uncoating the virus, thereby disabling it
- preventing newly-reproduced viral particles from constructing a protein coat
- preventing viral RNA from transcribing DNA (reverse transcriptase inhibitor)
- blocking newly-formed viral particles being released from the host cell (neuraminidase inhibitors).

 ISBN 978 0 6557 0026 5

PRACTICAL ACTIVITY 11

Enfuvirtide is an antiviral drug used to treat human immunodeficiency virus (HIV). It is an entry inhibitor, preventing HIV from entering T lymphocytes in the body. In order for HIV to bind to T lymphocytes, proteins on the outer coat of the virus must bind with proteins on the membrane of the host cell. Entry inhibitors attach themselves to proteins on the surface of either the virus or the cell, preventing HIV from gaining entry into cells.

Tamiflu and Relenza are two antiviral drugs registered for use in Australia that are used against influenza. They are in the class of antiviral drugs known as neuraminidase inhibitors. To be released from the cell, newly-formed viral particles use a protein on their surface, an enzyme called neuraminidase. Tamiflu and Relenza are drugs whose molecules bind to the viral enzyme neuraminidase, preventing virus particles from leaving the host cell.

AIM

- To simulate the action of selected chemical agents against pathogens.
- To consider some issues related to drug resistance by pathogens.

MATERIALS

- 10 students (plasma membrane) (adjust size of 'cell' to suit smaller classes)
- 15 black balloons (bacteria)
- 1 pin (antibiotic)

PART A • ANTIBIOTICS

METHOD

1. 10 students stand in a wide circle where they are able to touch hands with the person on either side to represent a cell, with the students as the plasma membrane.
2. Inflate one black balloon to represent a bacterium that has entered the cell and place it in the centre of the circle.
3. Bacteria reproduce by binary fission and under optimal conditions can multiply every 20 minutes. To simulate this, inflate one more balloon in 20 seconds and place it on the floor inside the cell.
4. Inflate two more balloons in the next 20 seconds. Place the balloons on the floor in the centre.
5. Inflate four more balloons in the next 20 seconds. Place the balloons on the floor in the centre.
6. Have one student act as a bactericidal antibiotic and proceed to pop the balloons with the pin.

1 Complete the table below showing the reproductive pattern for bacteria.

Time (hours)	0	1	2	3	4	5	6	7	8	9	10	11	12
Number of bacteria	1												

2 **a** If the bacteria in question can reproduce every hour, what will the bacterial population in the host be after 24 hours? (Clue: look for a pattern to help find the answer.)

b Use the grid shown to sketch a graph of the general population growth for the bacterium.

c Describe the population growth of the bacteria. How might this impact on the health of an infected individual?

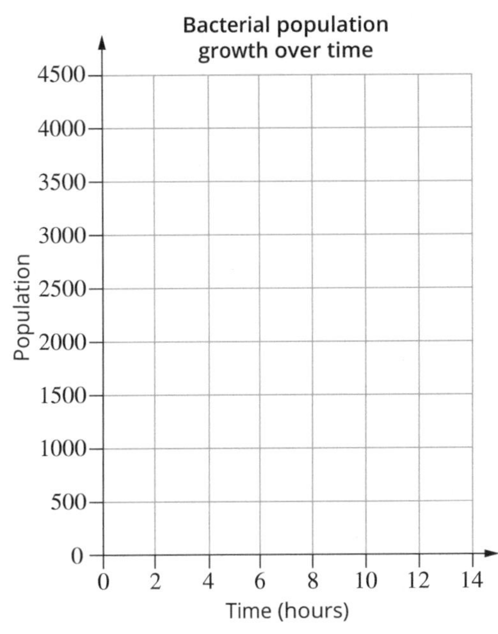

3 In what way does the popping of the balloons by the pin simulate the action of an antibiotic?

4 Is it possible for an antibiotic to damage 'good' bacteria in the body? Explain.

5 Why does an antibiotic affect the outside of the bacteria and not the bacterial cell itself?

6 What is the key difference between a bactericidal and bacteriostatic antibiotic?

PART B • ANTIVIRALS

Unlike bacteria, viruses are not cellular in structure. As such, antiviral drugs are strategically different to antibiotics—they target the features of viruses in different ways. Successful antivirals have been developed and are being developed that focus on the different strategies used by viruses to enter or leave cells or the way they operate inside cells, including interrupting processes such as transcription of RNA to DNA, viral protein synthesis and assembly of the viral coat.

Figure 4.1.13 illustrates the steps in the reproduction of a virus, from the virus binding to cell receptor protein (1), entering the cell (2), uncoating of the viral capsule (3), incorporation of viral RNA/DNA into host DNA (4), production of viral protein using cell's ribosomes (5), assembly of a new virus particle at the plasma membrane (6) and the release of the virus from the host cell (7).

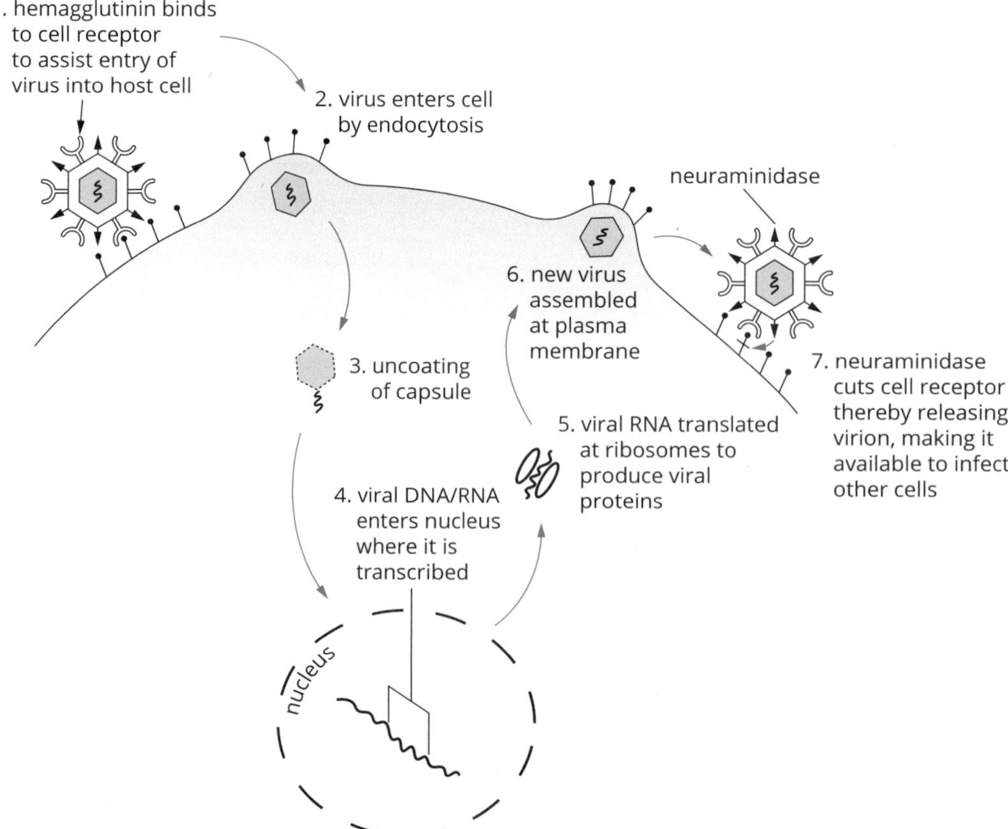

Figure 4.1.13 Viral entry into cell, reproduction and release

 ISBN 978 0 6557 0026 5

PRACTICAL ACTIVITY 11

7 Enfuvirtide is a drug used to inhibit HIV from infecting T lymphocytes.

a Complete the diagram below, showing where the entry inhibitor binds.

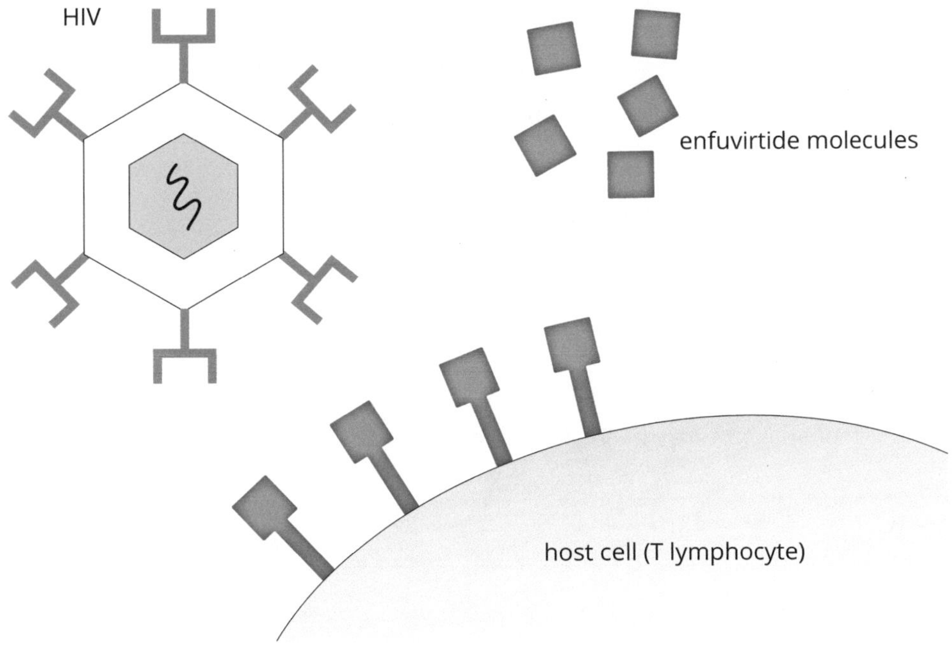

b Explain how the antiviral drug works against the virus.

8 Relenza is a neuraminidase inhibitor.

a On the diagram below, sketch in molecules of neuraminidase inhibitor where the drug would be attached to the virus.

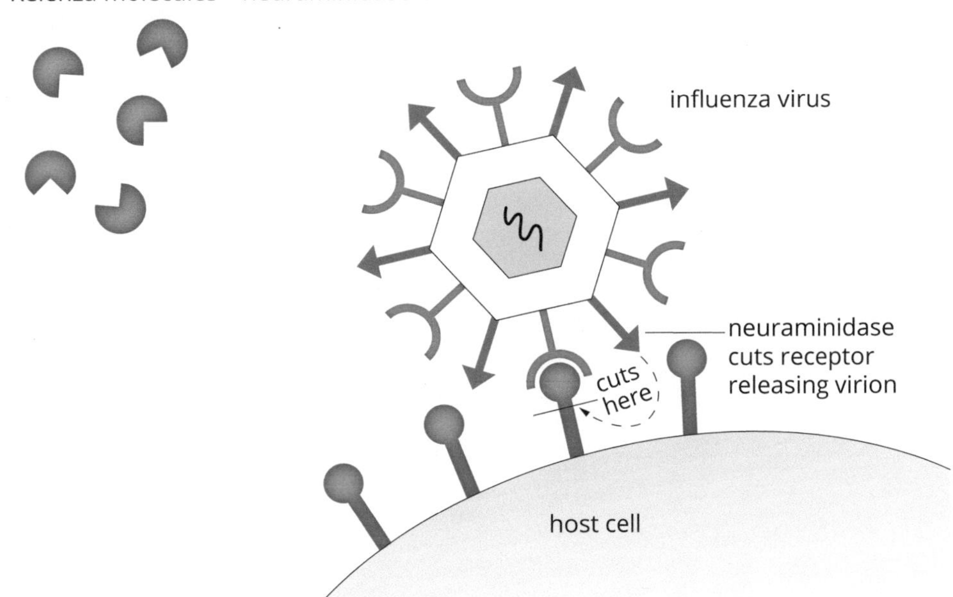

ISBN 978 0 6557 0026 5

b Explain how the antiviral drug works against the virus.

9 Explain whether neuraminidase inhibitors such as Relenza are likely to bind tightly and permanently to the neuraminidase active site, or loosely and temporarily.

10 Suggest why the use of antivirals may be limited after 48 hours of becoming sick with the flu.

11 One disadvantage of developing vaccines to protect the community against viruses such as influenza is that the variable part of virus particles can be altered. When this occurs, the antibodies that are specific to the original virus do not 'fit' any altered versions of the same virus. Outline how antiviral drug design could overcome this problem.

12 Antibiotic resistance in bacteria is an increasing problem in medicine.

a Explain what is meant by multi-drug resistance in some pathogenic bacteria.

b Describe how antibiotic resistance can develop with the overuse of antibiotics.

c Outline the potential consequences of drug resistant bacteria.

ISBN 978 0 6557 0026 5

CONCLUSIONS

13 Explain why it is important to take antibiotics and antivirals early in an infection.

14 Summarise the effectiveness of different kinds of antibiotics on bacterial infections.

15 Outline the similarities and differences between antibiotics and antivirals.

EXAM QUESTIONS

Multiple-choice questions

Question 1 VCE Biology 2018 (A) 16

Lupus is a condition that results in the increased secretion of antibodies that attach themselves to healthy cells in a patient's body. The accumulation of these antibodies causes inflammation, joint pain, rash, fatigue and fever.

Lupus is an example of

A. an allergic reaction.

B. an autoimmune disease.

C. an immune deficiency disease.

D. a complement protein response.

Question 2 VCE Biology 2018 (A) 17

Antigen-presenting cells deliver antigens to lymphocytes found within lymphoid tissue. These lymphocytes recognise these antigens as being non-self, causing the lymphocytes to become activated and increase in number.

Which of the following lymphocytes are activated by antigen-presenting cells?

A. helper T cells

B. dendritic cells

C. plasma cells

D. neutrophils

Question 3 VCE Biology 2018 (A) 23

Rabies is a viral disease spread to people by infected animals. A person bitten by an infected animal should be given an injection of specific antibodies.

Following the injection, this person should have

A. natural active immunity.

B. artificial active immunity.

C. natural passive immunity.

D. artificial passive immunity.

Question 4 VCE Biology 2018 (A) 24

Monoclonal antibodies can be produced and used to treat different types of cancers.

Which one of the following is a correct statement about monoclonal antibodies?

A. Monoclonal antibodies are carbohydrate molecules.

B. Monoclonal antibodies produced from the same clone of a cell are specific to the same antigen.

C. Monoclonal antibodies pass through the plasma membrane of a cancer cell and attach to an antigen within the cell.

D. Monoclonal antibodies produced to treat stomach cancer will be identical to monoclonal antibodies produced to treat breast cancer.

ISBN 978 0 6557 0026 5

EXAM QUESTIONS

Use the following information to answer questions 5 and 6. VCE Biology 2017 (A) 24 and 25

Multiple sclerosis (MS) is an autoimmune disease. In sufferers of MS, the myelin coating of nerve cell axons is damaged. This damage results in poor transmission of nerve messages between the brain, the spinal cord and the rest of the body. One aspect of MS diagnosis is imaging the brain to detect visible areas of demyelination, called plaques.

Question 5 VCE Biology 2017 (A) 24

Researchers investigating MS have analysed various tissue samples from patients.

In these samples they would expect to find

A. an abundance of allergens in nerve cells.

B. cancer cells in MS plaques in brain tissue.

C. increased numbers of helper T cells in spinal fluid.

D. an absence of T cytotoxic cells in the spinal cord and brain.

Question 6 VCE Biology 2017 (A) 25

Scientists are investigating factors that increase the likelihood of developing MS. Recently, the 'hygiene theory' has been considered a possible factor. This theory proposes that, if a child's environment is overly hygienic and does not allow sufficient exposure to a wide range of non-self antigens, an overactive immune system will result later in life.

A recent study tested for the presence of antibodies to the bacteria that cause stomach ulcers, *Helicobacter pylori*, in the blood of 550 MS patients and 299 healthy people. Both groups of people had the same proportion of each gender and were of similar age. Exposure to *H. pylori* usually occurs by the age of two years. The results of the antibody testing showed that the rate of *H. pylori* infection was 30% lower in the women with MS than in the healthy women or healthy men.

The findings of this study are consistent with the suggestion that

A. monoclonal antibodies could be used to treat MS.

B. males are affected by MS 30% more often than females.

C. suffering from a stomach ulcer is a common symptom of MS.

D. in females childhood exposure to *H. pylori* helps to protect against MS.

Question 7 VCE Biology 2016 (A) 24

In the search for a malaria vaccine, scientists have focused on a protein called circumsporozoite protein (CSP). CSP is secreted by the malaria parasite and is present on its surface.

For the vaccination to work, the scientists want CSP to act as

A. an antigen.

B. an allergen.

C. an antibody.

D. a complement protein.

EXAM QUESTIONS

Question 8 VCE Biology 2017 (A) 26

A daily blood sample was obtained from an individual who received a single vaccination against a particular strain of the influenza virus. The individual had no prior exposure to this strain of influenza. The graph below shows the concentration of antibodies present in the individual's blood for this strain of influenza over a period of 65 days.

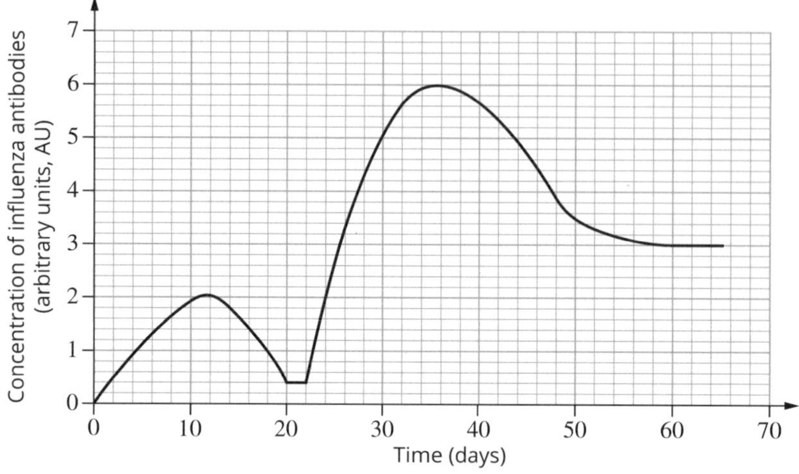

Which one of the following conclusions can be made using this data?

A. Memory B cells were activated by exposure to the same strain of the influenza virus on day 22.

B. B plasma cells specific to this strain of influenza were most numerous on day 12.

C. Herd immunity to this particular strain of influenza was achieved by day 55.

D. The vaccination containing weakened influenza antigens occurred on day 10.

Question 9 VCE Biology 2018 (A) 29

The graph below shows the death rates from acquired immune deficiency syndrome (AIDS) and also the number of people infected with the human immunodeficiency virus (HIV). Before 1995 many people infected with the virus went on to develop AIDS, which led to their deaths.

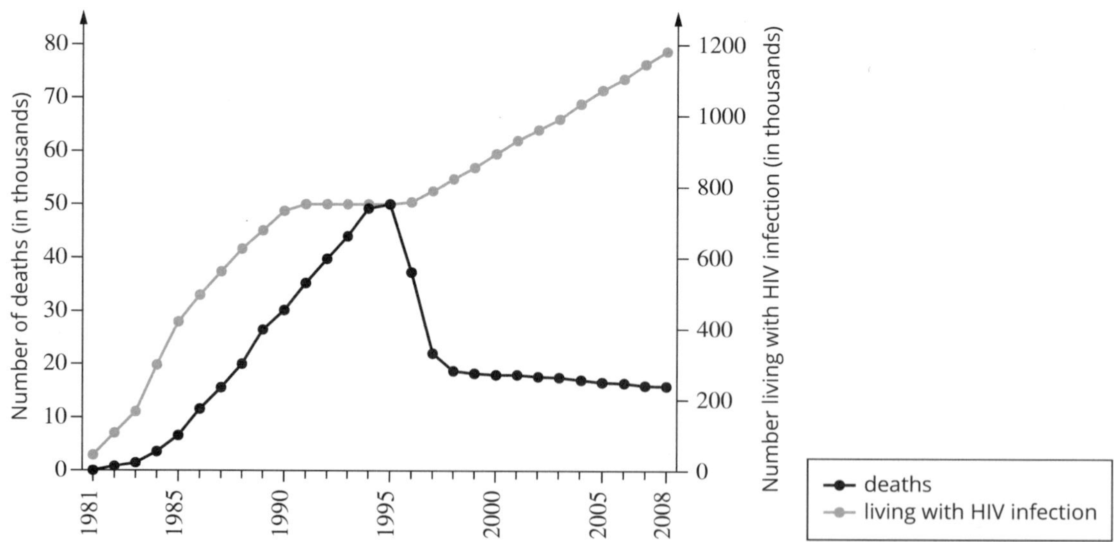

Based on the information in the graph, what is the most likely reason for the change in death rates, even though infection rates continued to climb after 1995?

A. People had access to new antiviral drugs.

B. People were educated about what caused HIV infection.

C. People infected with HIV were isolated from the rest of the public.

D. A widespread vaccination program for HIV was introduced within a targeted population.

ISBN 978 0 6557 0026 5

EXAM QUESTIONS

Question 10 VCE Biology 2018 (A) 35

The emergence of antibiotic-resistant diseases in humans means that

A. antibiotics are causing resistance mutations in bacteria.

B. some bacteria are less sensitive to antibiotics.

C. viruses are becoming resistant to antibiotics.

D. humans are less sensitive to antibiotics.

Short-answer questions

Question 1 (6 marks) VCE Biology 2018 (B) 3

Plants are a rich source of nutrients for many organisms, including bacteria, fungi and viruses. Although plants lack an immune system that is comparable to animals, plants have evolved chemical barriers to stop invading pathogens from causing significant damage.

a. Describe **two** chemical barriers that could be present in a plant that is protecting itself from an invading pathogen. 2 marks

b. Humans have a sophisticated immune response to invading pathogens. State **two** ways that pathogens are prevented from entering the internal environment of the human body. 2 marks

c. Outline how complement proteins and natural killer cells protect the human body once a pathogen has gained entry to the internal environment. 2 marks

Question 2 (11 marks) VCE Biology 2017 (B) 4

Australian marsupials, such as wallabies, kangaroos, wombats and koalas, give birth to very undeveloped young called joeys. When a joey enters the mother's pouch, it is at a stage equivalent to a seven-week-old human fetus. It spends many weeks in the pouch feeding on milk produced by mammary glands. Although the pouch provides protection from predators, it is neither sealed nor sterile.

a. **i.** What is meant by the term 'sterile' in the context given? 1 mark

ii. Consider a hospital environment. Give **two** examples of how sterile conditions can be achieved in a hospital. 2 marks

EXAM QUESTIONS

The joey's primary immune tissue (in the bone marrow and thymus) does not mature until 30 days after birth and its humoral immunity does not function effectively until 90 days after birth. Biologists have analysed milk samples from several marsupial species and found that they contain various antibodies. Some of the antibodies in the mother's milk remain in the joey's gut, while others cross the gut wall and enter the joey's bloodstream.

b. Describe at a molecular level how antibodies perform their function. 2 marks

c. Name the type of immunity that the joey obtains from the antibodies in the milk and explain how this form of immunity is beneficial to the joey. 3 marks

Scientists have found that the milk of the tammar wallaby (*Macropus eugenii*) contains high levels of peptides with antibiotic properties, as well as lysozyme, complement proteins, cytokines and venom inhibitors.

d. i. Name the part of the immune system to which these peptides and the other listed chemicals belong. 1 mark

ii. Circle one of the chemicals below, found in tammar wallaby milk, and describe its role in protecting the joey against pathogens. 1 mark

lysozyme complement proteins cytokines venom inhibitors

Role

e. Scientists tested the tammar wallaby milk peptides and found them to be 10 times more effective than antibiotics such as tetracycline and ampicillin, which are commonly used to fight human diseases. The scientists are keen to find a pharmaceutical company that will support further testing and development of these peptides with antibiotic properties. 1 mark

Suggest what the scientists would hope to achieve as a result of further testing of these peptides with antibiotic properties.

ISBN 978 0 6557 0026 5

EXAM QUESTIONS

Question 3 (7 marks) VCE Biology 2014 (B) 4

a. Define the term 'pathogen'. 1 mark

The diagram below shows a generalised pathogen with antigens on its surface. The immune system responds to antigens by making antibodies.

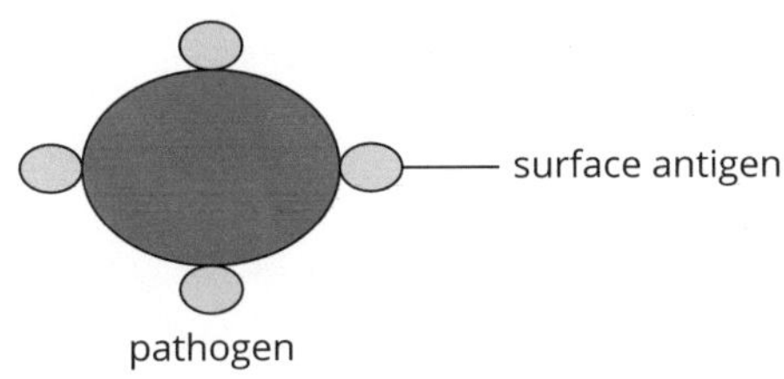

b. Draw an antibody that would be effective against this pathogen. Label the different parts of the antibody. 2 marks

One way that antibodies work is by forming antigen–antibody complexes.

The diagram below shows four pathogens.

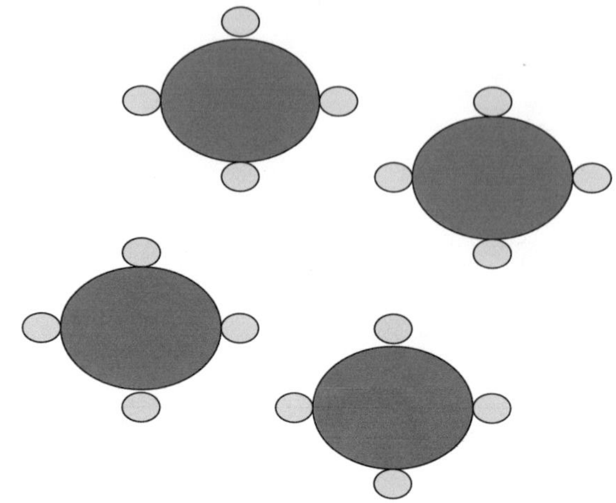

c. **i.** Illustrate on the diagram **above** how the antigen–antibody complex forms, using at least **four** antibodies in your drawing. 2 marks

ii. Explain how an antigen–antibody complex provides protection against these pathogens. 2 marks

Question 4 (7 marks) VCE Biology 2010(1) (B) 7

Coeliac disease in humans is caused when cells of the immune system attack the epithelial cells that line the small intestine.

a. What is the general name given to this type of disorder? 1 mark

Coeliac sufferers are unable to break down the gluten found in grains such as wheat. One of the features of celiac disease is 'leaky gut syndrome'. A small gap appears between the epithelial cells that line the small intestine. Gluten fragments enter the gap and accumulate under the epithelial cells. Macrophages are stimulated to remove the fragments.

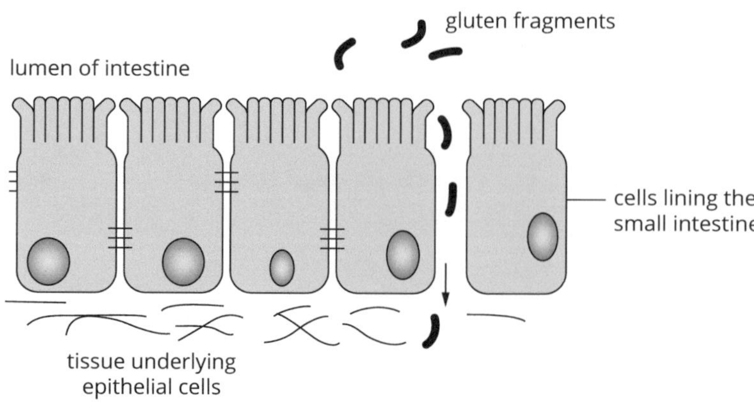

 ISBN 978 0 6557 0026 5

b. Explain how a macrophage is able to remove and destroy a gluten fragment. You may use a written answer or labelled diagrams or both. 2 marks

Once a macrophage has destroyed a gluten fragment, it displays a piece of the fragment on its membrane using a special major histocompatibility complex (MHC) marker. A helper T cell then attaches to the MHC marker antigen complex.

The macrophage helper T cell complex is shown below.

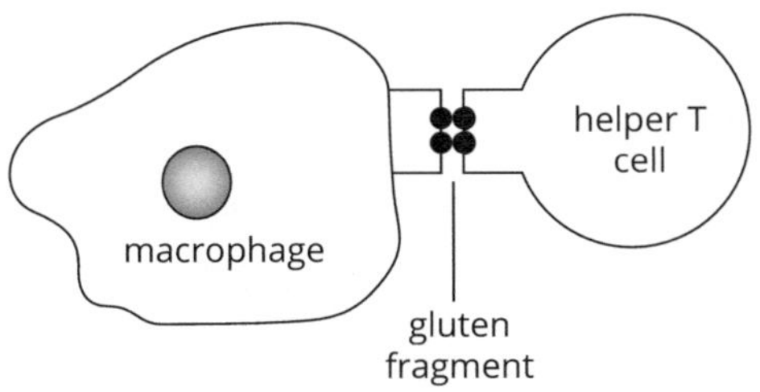

The macrophage helper T cell complex stimulates other cells and chemicals to target and damage epithelial cells that line the intestine.

c. Name one cell or chemical that would be stimulated by the macrophage helper T cell complex and state its function. 2 marks

Name

Function

Assume that a drug has been shown to be safe for human use and is about to be trialled. Assume that you suffer from coeliac disease and have been invited to take this drug as part of the trial.

d. What are two different questions about the drug trial that you would ask your doctor? 2 marks

Question 1 ____

____

____

Question 2 ____

____

____

Question 5 (9 marks) VCE Biology 2019 (B) 9

Zika fever is a rapidly emerging viral disease. It is most commonly transferred from one person to another by the *Aedes* species of mosquito.

Zika fever in people was discovered in Uganda in 1947. It was thought that a bite from a mosquito had transferred the virus from monkeys to humans.

The symptoms of Zika fever are usually mild and 80% of infected humans do not show symptoms. Infection of pregnant women, however, can cause severe defects in their babies.

a. One way that diseases, such as Zika fever, are thought to occur is when a pathogen infects humans from an animal host. 1 mark

Identify **one** social or economic factor that could lead to this transfer between hosts.

____

____

____

b. When scientists attempt to identify a particular disease, they can look for specific antibodies in infected humans. Scientists trying to identify Zika fever infections found that testing for the antibodies produced against the Zika virus often gave them incorrect results. This was because the antibody tests that had been developed could not always identify the difference between the antibodies produced against the Zika virus and the antibodies produced against other viruses. 3 marks

Explain why making a correct identification of a viral pathogen is important in the control of a disease.

____

____

____

____

c. Explain why the antibody tests could not identify the difference between the antibodies produced against the Zika virus and the antibodies produced against other viruses. In your response, refer to the structure of the antibody.

____

____

____

ISBN 978 0 6557 0026 5

EXAM QUESTIONS

d. *Aedes* mosquitoes are not found on every continent. They cannot fly great distances. Vaccines are currently being trialled for the Zika virus. 3 marks

Describe **three** different approaches, other than vaccination, that government health officials could use to reduce the spread of the Zika virus.

UNIT

4 How does life change and respond to challenges?

AREA OF STUDY 2

How are species related over time?

Outcome 2

On completion of this unit the student should be able to analyse the evidence for genetic changes in populations and changes in species over time, analyse the evidence for relatedness between species, and evaluate the evidence for human change over time.

Key knowledge

Genetic changes in a population over time

- causes of changing allele frequencies in a population's gene pool, including environmental selection pressures, genetic drift and gene flow; and mutations as the source of new alleles
- biological consequences of changing allele frequencies in terms of increased and decreased genetic diversity
- manipulation of gene pools through selective breeding programs
- consequences of bacterial resistance and viral antigenic drift and shift in terms of ongoing challenges for treatment strategies and vaccination against pathogens

Changes in species over time

- changes in species over geological time as evidenced from the fossil record: faunal (fossil) succession, index and transitional fossils, relative and absolute dating of fossils
- evidence of speciation as a consequence of isolation and genetic divergence, including Galápagos finches as an example of allopatric speciation and *Howea* palms on Lord Howe Island as an example of sympatric speciation

Determining the relatedness of species

- evidence of relatedness between species: structural morphology—homologous and vestigial structures; and molecular homology—DNA and amino acid sequences
- the use and interpretation of phylogenetic trees as evidence for the relatedness between species

Human change over time

- the shared characteristics that define mammals, primates, hominoids and hominins
- evidence for major trends in hominin evolution from the genus *Australopithecus* to the genus *Homo*: changes in brain size and limb structure
- the human fossil record as an example of a classification scheme that is open to differing interpretations that are contested, refined or replaced when challenged by new evidence, including evidence for interbreeding between *Homo sapiens* and *Homo neanderthalensis* and evidence of new putative *Homo* species
- ways of using fossil and DNA evidence (mtDNA and whole genomes) to explain the migration of modern human populations around the world, including the migration of Aboriginal and Torres Strait Islander populations and their connection to Country and Place.

KEY KNOWLEDGE

Genetic changes in a population over time

A change in the genetic make-up of populations over time is called **evolution**. These changes are manifested in the observable **phenotypes** (traits) of organisms. Changes in the proportions of phenotypes present in populations occur as a consequence of changes in the frequency of **alleles** over generations. Phenotypes are subject to environmental pressures. Allele frequencies within populations change as a result of environmental pressures that act on those phenotypes.

ALLELE FREQUENCIES

- **Gene pool**: the genetic make-up of a population; includes the sum of all of the alternative alleles for different genes present in a population.
- Allele: a variant of a gene.
- **Genotype**: the genetic make-up of an individual—the particular alleles present for a gene.
  - **Homozygotes** have two copies of the same allele for one gene (e.g. *AA* or *aa*).
  - **Heterozygotes** have two different alleles for one gene (e.g. *Aa*).
- **Allele frequency**: the proportion of a particular allele in a population.
  - Allele frequencies range between 0 and 1.
  - An allele frequency of 0 means that no individuals in the population have the allele.
  - An allele frequency of 1 means that all individuals in the population have the allele and are homozygous for that allele.
  - The sum of the alternative alleles for a given gene adds up to 1.

The following equation shows you how to calculate the frequency of an allele.

$$\text{allele frequency} = \frac{2(\text{number of homozygotes}) + (\text{number of heterozygotes})}{2(\text{total number of individuals})} \times 100$$

Formula:

$$p = \text{frequency of dominant allele}$$
$$q = \text{frequency of recessive allele}$$
$$p + q = 1$$

Example: In a population of 10 kookaburras, there are nine individuals with normal feather pigmentation and one albino. Of the nine individuals with normal colouring, six are homozygous dominant and three are heterozygous.

N = allele for pigmented feathers

n = allele for no pigment (albino)

10 kookaburras = 20 alleles for feather pigment (each individual has 2 alleles for this gene)

Six homozygous dominant birds ($2 \times 6N$) + three heterozygous birds ($3N$) = 15 N alleles, i.e. frequency of dominant allele (p) = 15/20 = 0.75

One homozygous recessive (albino) bird ($2n$) + three heterozygous birds ($3n$) = 5 n alleles in population, i.e. frequency of recessive allele (q) = 5/20 = 0.25

$p + q = 1$

$0.75 + 0.25 = 1$

The equation is useful for determining the frequency of an unknown allele when the frequency of the other allele is known.

Selection pressures are factors in the environment that act on the phenotypes of individuals. As a result, the frequency of alleles in the population may change.

Artificial selection occurs when humans intervene in the breeding of organisms such as plants and animals to deliberately select for breeding purposes those individuals with the most desirable characteristics. Such selective breeding programs alter the gene pool for these populations.

VARIATION

Populations show variation in the forms of traits. A trait that has many forms is called **polymorphic**.

Genetic diversity in populations is generated as a result of sexual reproduction and mutation.

Sexual reproduction involves the production of genetically unique gametes in meiosis resulting from independent assortment of chromosomes and genetic recombination in crossing over. The random union of gametes in fertilisation further mixes the genetic material.

Mutation is the raw material for evolutionary change. Mutations introduce new alleles into populations.

ISBN 978 0 6557 0026 5

KEY KNOWLEDGE

MUTATION

A mutation is a change in the genetic material of an organism, that is, a change to the sequence of bases in the DNA. Mutations can be spontaneous or induced. Spontaneous mutations are naturally occurring random changes in genetic material; they occur for no apparent reason. Induced mutations are changes in the genetic material that occur as a result of exposure to a mutagenic agent—an agent that causes a change in the DNA base sequence (e.g. radiation).

Most mutations are detected and repaired by enzymes. Those that cannot be repaired fall into one of three categories—neutral, beneficial or harmful.

- Neutral mutations have no effect on survival.
- Beneficial mutations increase the likelihood of survival.
- Harmful mutations decrease the likelihood of survival.

Somatic mutations occur in somatic cells. They only affect the individual in which they occur and are not passed on to the next generation.

Germline mutations occur in the cells that give rise to gametes and are passed on to individuals in the next generation.

POPULATION CHANGE AND ALLELE FREQUENCIES

Populations are subject to change as a result of various factors—gene flow, genetic drift, the founder effect and genetic bottlenecks.

- **Gene flow**: the exchange of alleles between populations as a result of migrating individuals who reproduce in the new populations they join. Gene flow in plants occurs through the dispersal of pollen and seed. Gene flow can introduce new alleles into a population, changing its genetic composition.

  Example: The movement of Fy^{o}, Fy^{a} and Fy^{b} alleles of the Duffy blood group in humans between populations of African and European origin. The European settler population in the United States showed a high frequency of the Fy^{a} allele and the African population a low frequency. Today, gene flow accounts for an overall increase of this allele in African Americans.

- **Genetic drift**: change in allele frequencies in a population over generations as a result of chance alone. That is, no selection pressures appear to be acting on the phenotypes present. Small, isolated populations are more susceptible to genetic drift than larger populations or populations that maintain gene flow with neighbouring populations.

  Example: A new mutation that is 'neutral', that is, does not put the individual at a survival disadvantage, may become fixed in a population, changing the allele frequencies in the population.

  Genetic drift can occur when populations decrease for a period of time (a bottleneck effect) or in small founding populations (the founder effect).

  - **Founder effect**: occurs when a small population of individuals is isolated and forms the basis of a new population. A founder group of individuals may have limited variation, and not be representative of the population from which it originated. After successive generations, the population is phenotypically different from its parent population.
  - **Bottleneck effect**: occurs when a large population is dramatically reduced in size, thereby reducing its genetic diversity. The frequency of some alleles in the population may thus be reduced or eliminated altogether. Such events may reduce the chances of survival of a population or species, especially if favourable phenotypes are poorly represented in the remaining population.

    Examples: Catastrophic events such as drought and volcanic activity can dramatically reduce population sizes. A meteorite strike is one theory that has been proposed to explain the bottleneck that occurred in the dinosaur population approximately 65 million years ago, eventually leading to their extinction.

    The helmeted honeyeater and Leadbeater's possum are examples of Australian species that have experienced genetic bottlenecks following the loss of habitat. Both these species are currently listed as Critically Endangered.

Changing allele frequencies can have significant consequences for populations and species. When new alleles are introduced into populations by gene flow from neighbouring populations of the same species or by mutation, there is an increase in genetic diversity, which can be advantageous, particularly in a changing environment with new selection pressures. When alleles are removed from a population, its genetic diversity decreases. Low genetic diversity makes populations more vulnerable, especially when environmental conditions are altered.

Macroevolutionary change

Macroevolution refers to large-scale changes in the evolution of organisms that occur above the species level. Allele frequencies change in populations over many generations leading to cumulative changes that can result in speciation, that is, new species diverge from ancestral ones.

Example: Modern species of Australian marsupials that have diverged from ancestral marsupial species.

KEY KNOWLEDGE

Microevolutionary change

Microevolution refers to small-scale changes within species, such as the differences in allele frequencies of populations from one generation to the next.

Examples: The frequency of alleles conferring antibiotic resistance is increased from one generation of bacteria to the next; the frequency of alleles conferring resistance to myxomatosis is higher in subsequent generations of a rabbit population.

Antigenic drift is an example of microevolutionary change in viruses and refers to the gradual accumulation of mutations in viral DNA or RNA. These genetic changes translate to changes in the antigens present on the protein coat of a virus. Usually, the changes in antigen structure are minor and the antibodies in the host's immune system are able to recognise the virus and mount an immune response if a similar virus has infected the host before. However, over time, multiple episodes of antigenic drift can result in significant changes to viral antigens, effectively creating a new virus that the host's immune system no longer recognises.

Antigenic shift is a much more abrupt change in the genome of a virus and results from the re-assortment of genetic material from different viral strains. Gene swapping between viral strains can result in a virus with different antigens and new characteristics, such as increased virulence and the ability to infect new hosts. Antigenic shift typically occurs when an individual is infected with two different viruses at the same time, providing opportunity for the viruses to swap genetic material in the host's cells. When a virus that usually infects a non-human animal species acquires the ability to infect humans and cause disease, it is known as a **zoonotic disease**. Many cases of antigenic shift are zoonotic in origin. COVID-19 is an example of a such a disease and is extremely contagious and potentially fatal. New infectious diseases, such as COVID-19, present unique challenges for scientists. Strategies for controlling the spread and transmission of new pathogens usually include a combination of treatments, health education campaigns, behavioural changes (such as social distancing and isolation) and the development of new vaccines.

NATURAL SELECTION

Natural selection is a process in which individuals with the most favourable characteristics have a better chance of surviving and reproducing than individuals with less favourable characteristics. The ability of an organism to survive and produce viable offspring is known as biological fitness (Table 4.2.1).

An **adaptation** is a feature displayed by an organism that makes it well suited to its particular environment or lifestyle. For example, wings are an adaptation for flight, and colour can provide camouflage—an adaptation to avoid predation.

Table 4.2.1 Change in population over time and generations

variation exists within populations; some individuals are more suited (adapted) to the environment, others are less well adapted	
↓	↓
individuals that are most adapted are said to have the most favourable characteristics	individuals with less favourable characteristics are less well adapted
↓	↓
individuals have a survival advantage and so have an increased chance of reproducing and passing on their 'favourable' alleles	individuals have a decreased chance of surviving and reproducing compared to those with favourable characteristics
↓	↓
they leave a greater number of offspring that also have favourable alleles, i.e. offspring adapted to environment	leave fewer offspring
↓	↓
increase in frequency of favourable alleles in population	decrease in frequency of 'unfavourable' alleles in population
↓	↓
overall increase in proportion of individuals with favourable characteristics in population, and overall decrease in proportion of individuals with less favourable characteristics	
↓	↓
population becomes increasingly adapted to its environment	

Changes in species over time

Species have undergone change as long as life has existed on Earth. A look at the fossil record reveals evidence that life on Earth has become more diverse and more complex over geological time.

THE FOSSIL RECORD

Evolution is the process of change in organisms over time. Fossils provide strong evidence supporting the theory of evolution. A **fossil** is the remains or trace of an organism that lived before the present time. Fossils exist in different forms.

- **Preserved remains**: all or part of an organism remains intact (e.g. mineralised bones, shells, insects trapped in tree resin, mammoths preserved in ice sheets).
- **Impression fossils**: remains of organisms become encased in rock; the remains dissolve but an imprint of the organism is left in the rock (e.g. leaf impressions in rock).

 ISBN 978 0 6557 0026 5

- **Cast fossils**: remains of organisms encased in rock dissolve away and the space left becomes filled with foreign material, for example silt that forms a 3D replica (sculpture) of the organism (e.g. Gogo fish in the Kimberley, Western Australia).
- **Trace fossils**: indirect evidence that an organism existed (e.g. footprints, nests and scats).

Fossilisation of organisms occurs most effectively when certain conditions are met. These include a lack of oxygen, which reduces the rate of decay by bacteria, and a lack of scavengers. For these reasons, aquatic organisms are well represented in the fossil record (Table 4.2.2).

Table 4.2.2 Fossilisation of aquatic organisms

fish dies and sinks to sea floor
↓
water currents cover dead remains with sand before scavengers reach it
↓
depth of sea floor and burial by sand reduce oxygen available for decomposition by bacteria
↓
weight of water and sediments apply pressure to dead remains, replacing hard tissue such as bones with minerals

The **fossil record** indicates that:

- some algae and primitive animal life forms have existed on Earth for over 600 million years, with evidence of prokaryotic cells as old as 2500 years
- there has been an increase in the complexity of life forms over time
- there has been an increase in the **biodiversity** (variety of life forms) over time
- extinction of species has also occurred and continues to occur. Mass extinctions are extinction events characterised by the disappearance of a large number of species in a relatively small timeframe. Five mass extinction events are recognised in the fossil record (Figure 4.2.1).

Dating techniques and the geological time scale

The fossil record shows the emergence of different kinds of living organisms set against geological time. Geological time is a measure of the history of Earth in terms of its geology. The **geological time scale** is divided into five eras:

- Archaean: oldest rock 4000–2500 million years old
- Proterozoic: 2500–541 million years old
- Palaeozoic: 541–252 million years old
- Mesozoic: 252–66 million years old
- Cenozoic: 66 million years–present time.

The earliest known fossils have been dated as Precambrian and include stromatolites (prokaryotes) and multicellular Ediacaran fauna. Trilobites are examples of early Palaeozoic life forms.

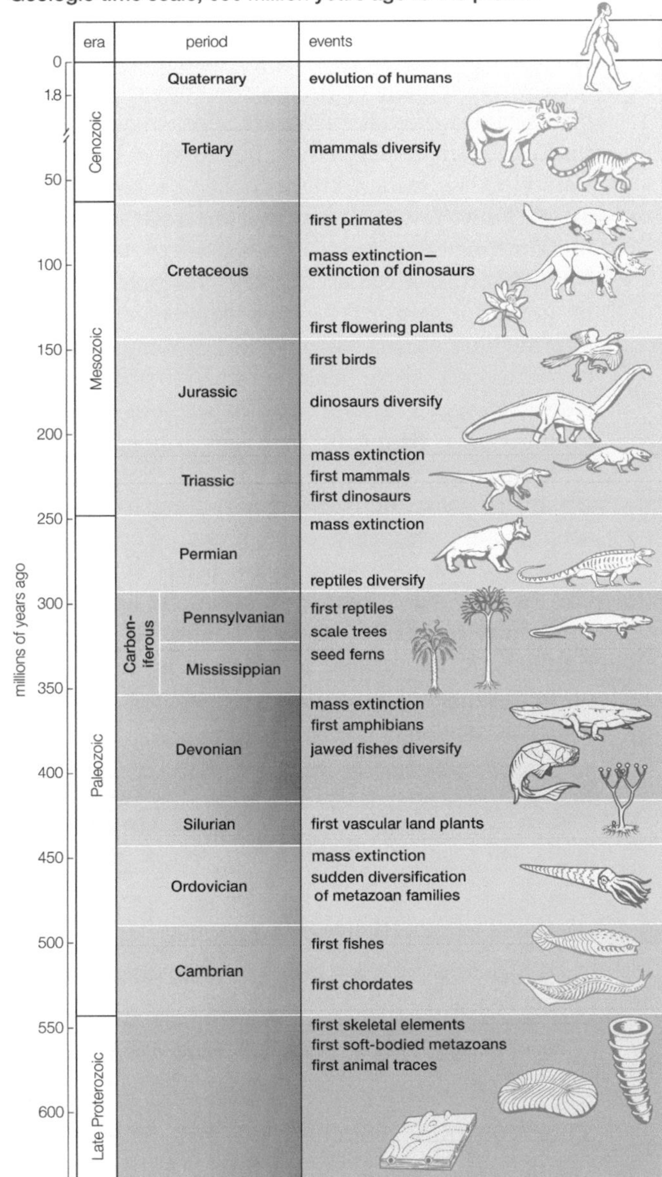

Figure 4.2.1 Geological time scale, 650 million years to the present, illustrating an increase in the complexity and diversity of life forms on Earth, including the rise of flowering plants and mammals

Various dating techniques are available to estimate the relative and actual age of rock and fossils.

- **Stratigraphy**: comparison of rock strata and the fossils they contain. The earliest rock strata were laid down first—these represent the oldest rock; upper layers are more recent. The relative age of a fossil can be determined by comparing the position of the rock stratum in which it was found with rock strata above and below. It follows that the lowest layers contain the oldest fossils and the highest layers contain the most recent fossils. This succession of fossils provides information about the relative ages of fossils, giving insights into the kinds of organisms that lived at different times in geological history.

KEY KNOWLEDGE

- **Index fossils**: useful in dating unidentified fossils. Index fossils are fossils that have been identified; are common in rock strata of a particular age; exist in rock where the age of the rock is known in at least one location. Fossils discovered in a new location that are either similar to the index fossil or have been found in association with the fossils of the same kind as the index fossil suggest that the rock stratum is the same age.
- **Transitional fossils**: display features present in both ancestral and descendant groups. Such intermediate forms suggest an evolutionary link between one form and another. *Archaeopteryx* is considered a transitional fossil suggesting an evolutionary pathway between dinosaurs and modern birds.
- **Absolute dating** techniques: determine with accuracy the actual age of rock or fossils. Absolute dating techniques measure the decay rates (half-life) of radioactive **isotopes** to assess age. This is called **radiometric dating** (Table 4.2.3). The **half-life** of an isotope is the time taken for half of the atoms in the radioactive sample to decay (undergo chemical change).

Table 4.2.3 Radiometric dating technique

Radiometric technique	Isotopes	Dating limitations
radiocarbon dating	measures ratio of C^{14} to C^{12}	less than 60 000 years old
potassium–argon dating	K^{40} : Ar^{40}	20 000 years–4600 million years old
uranium–lead dating	U^{238} : Pb^{207}	1 million years–4600 million years old

- **Thermoluminescence**: used to date human artefacts such as cooking utensils. It measures light emissions from minerals when heated. The intensity of light emitted from such objects provides a measure of the amount of time that has passed since the object was heated.

SPECIATION

A group of organisms is considered a distinct **species** if its members are similar and are able to interbreed in their natural environment to produce **viable offspring**.

Speciation is defined as the development of new species from pre-existing species.

Speciation occurs when populations become isolated from one another and no gene flow occurs between them. The different populations are subject to different selection pressures. Over time and many generations, the populations diverge, becoming so different that individuals are no longer able to interbreed, even if they do come together again.

Several models account for the evolution of new species from ancestral species. Each takes into account the factors acting on populations.

- **Allopatric speciation**: the most common form of speciation; occurs when a geographical barrier such as an ocean, desert, a mountain range or glacier divides a population. The separation of once-united landmasses due to movement of the Earth's crustal plates (continental drift) is another example of a geographical barrier that can lead to allopatric speciation. Populations are physically separated from one another, so gene flow between the populations ceases, resulting in reproductive isolation. Selection pressures act on the phenotypes in the separated populations. Changes in the populations may eventually lead to the development of distinct species. The process is gradual, occurring over a great deal of time and over many generations (Figure 4.2.2). Isolation of finch populations in the Galápagos Islands is an example of allopatric speciation. Other notable examples of allopatric speciation include the various species of flightless birds on different continents. These species evolved from a common ancestral species that inhabited a common landmass; separation of the landmasses resulted in populations isolated by expanses of ocean. Eventually speciation resulted in the similar but different species we see today. Many of Australia's unique species have evolved following geographical isolation and allopatric speciation.
- **Sympatric speciation**: occurs when gene flow is disrupted between two groups within the same geographic area; usually these are populations inhabiting different microhabitats within the same larger geographic area. Eventually the population change results in separate species. The *Howea* palms of Lord Howe Island are widely accepted as classic examples of sympatric speciation. Both *Howea belmoreana* and the less abundant *Howea forsteriana* are believed to have evolved on the volcanic Pacific island from a common ancestral species. Speciation is generally gradual, however when the right factors coincide, evolutionary change can occur over relatively short periods of time.
- Rapid speciation: may occur when a small, non-representative group becomes isolated from the original population. Advantageous mutations become fixed in the population and rapidly increase in frequency. The new group becomes different from the parent population relatively quickly, leading to speciation over a much briefer period of time.
- **Adaptive radiation**: a form of speciation in which an ancestral population is divided into a number of populations that diverge to form different species. The 13 finch species of the Galápagos Islands are a cluster of related species that represent an example of both allopatric speciation and adaptive radiation (Figure 4.2.3 on page 150).

ISBN 978 0 6557 0026 5

KEY KNOWLEDGE

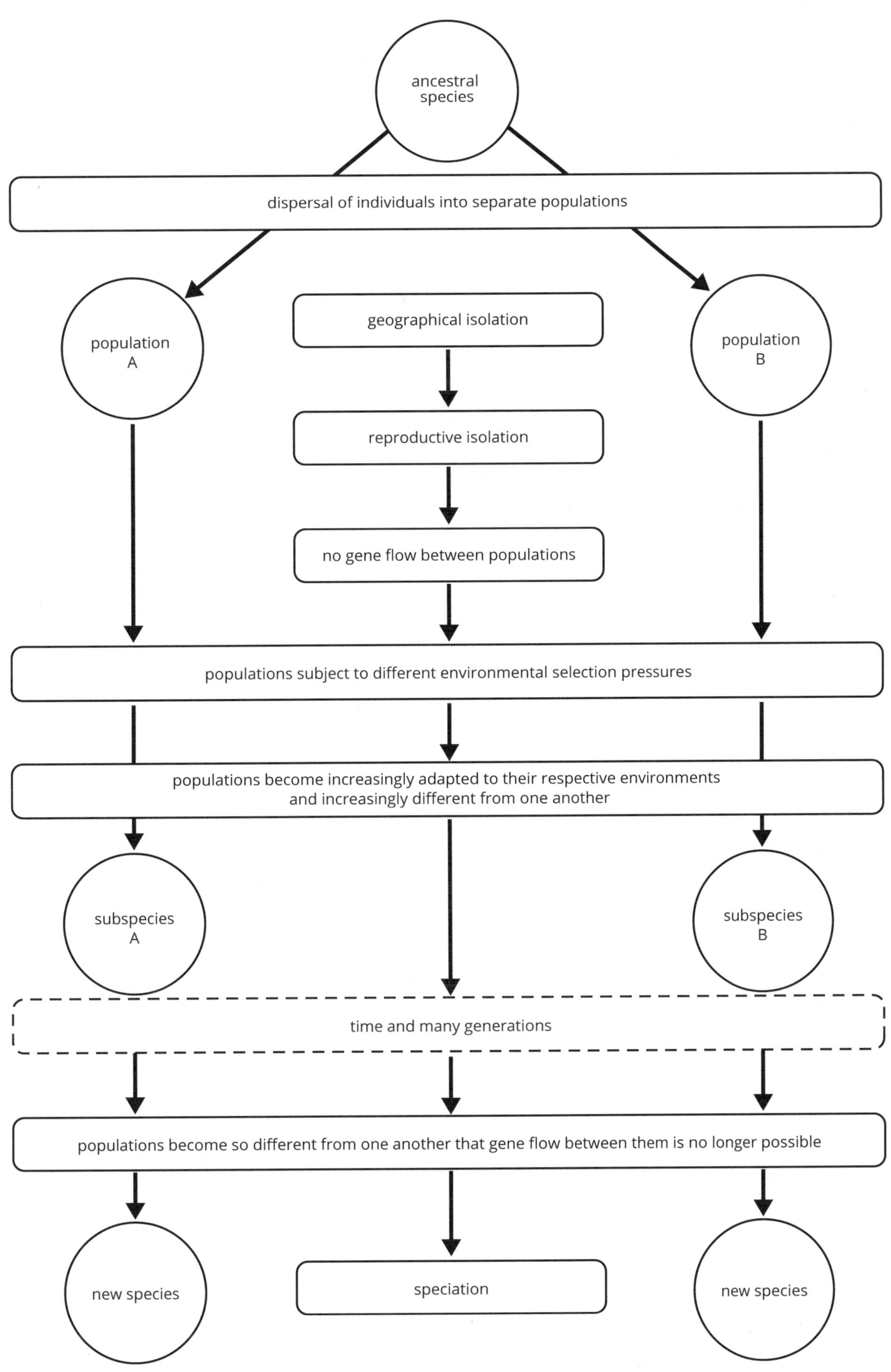

Figure 4.2.2 Model of speciation

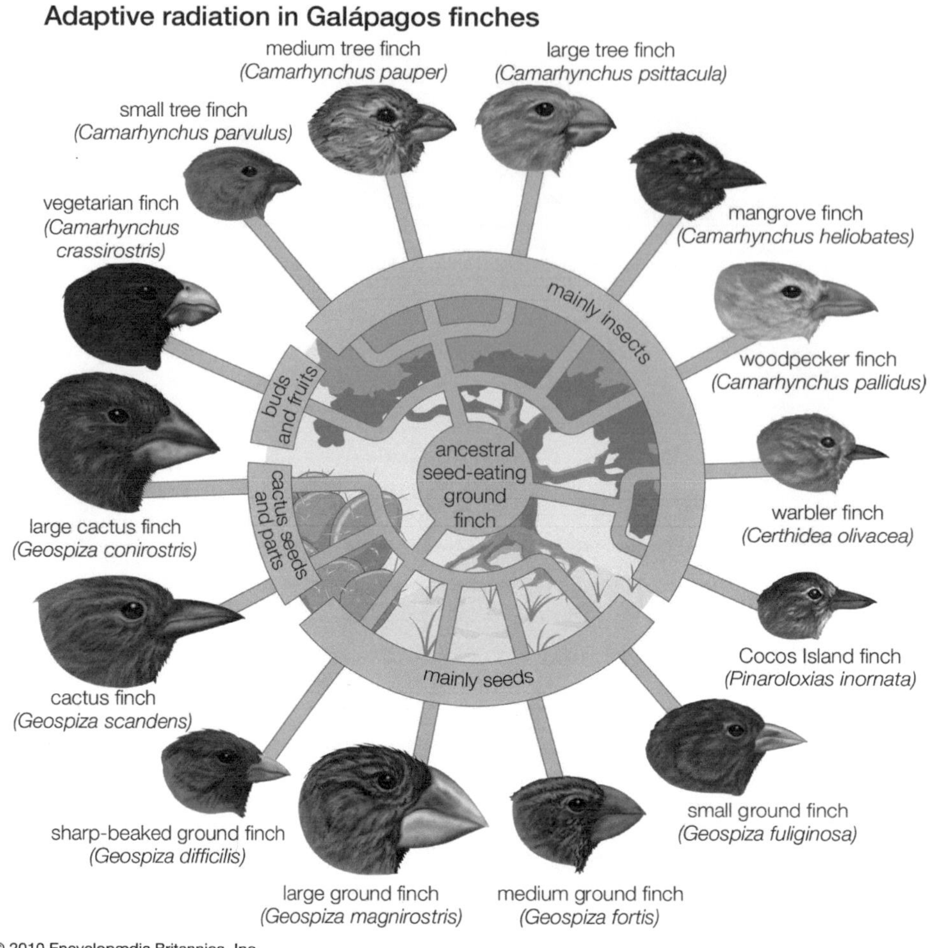

Figure 4.2.3 Adaptive radiation of Darwin's Galápagos finches. These separate species evolved from a common ancestor arriving on the Galápagos Islands.

ISOLATING MECHANISMS BETWEEN SPECIES

Similar, related species that share a particular habitat do not interbreed in their natural environment. **Genetic isolation** (also called reproductive isolation) occurs when alleles are no longer exchanged between populations. The mechanisms of genetic isolation are classified into two main types:

- before reproduction—**prezygotic isolating mechanisms**
- after reproduction—**postzygotic isolating mechanisms**.

Prezygotic isolating mechanisms

- **Geographical isolation**: physical and geographical barriers such as oceans, deserts, mountain ranges and glaciers separate populations.
- **Ecological isolation**: where species have different requirements that can only be met in different habitats. These may be in relatively close proximity, for example, different species of eucalypt growing on a mountain slope but one at a higher altitude than the other.
- **Temporal isolation**: the breeding cycles or active times of populations do not overlap.
- **Behavioural isolation**: the behaviours of members of different species mean that they do not recognise one another for the purposes of mating, or they do not come into contact for other reasons. For example, related species of frogs may have different mating calls that only members of the same species recognise; one species may be nocturnal while another related species is diurnal—their daily activity routines do not coincide.
- **Morphological isolation**: structural differences in the genitalia of the two species in question makes interbreeding impossible.
- **Gamete mortality**: egg and sperm are unable to fuse in fertilisation.

Postzygotic isolating mechanisms

- **Hybrid inviability**: the sperm from one species successfully fertilises the egg of another species to form a hybrid zygote, but the hybrid zygote has unmatched (non-homologous) chromosomes and normal embryonic development cannot proceed.
- **Hybrid sterility**: mating between species occurs and offspring are able to develop but the offspring are infertile and no further offspring can be produced.

 ISBN 978 0 6557 0026 5

KEY KNOWLEDGE

Determining the relatedness of species

While there are obvious differences between the different species of organisms on Earth, there are also remarkable similarities. Scientists analyse the similarities and differences between various species using a range of techniques to understand how closely related they are in evolutionary terms.

STRUCTURAL MORPHOLOGY

Structural morphology (also called comparative anatomy) examines the structural similarities and differences between different groups of organisms. It poses the question: how is it that different kinds of organisms possess similar basic anatomy?

Observation: the limbs of quite different groups of tetrapods (four-limbed animals) display a similar or homologous skeletal structure (Figure 4.2.4).

Conclusion: the different groups of organisms are related and share a common ancestor, from which these different groups diverged.

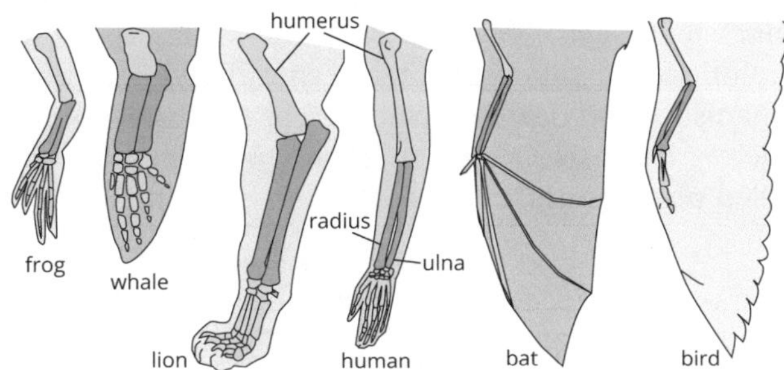

Figure 4.2.4 The bones in the forelimbs of tetrapods are homologous.

- **Homologous features**: fundamentally similar in morphology (for example, the wings of bats, hands of humans and flippers of seals), and show similar basic bone structure and arrangement, but have different proportions related to different functions. Homologous structures are indicative of common evolutionary ancestry.
- **Vestigial structures**: appear in reduced form in one species compared to a more developed form in another, for example, the human appendix is the same anatomical structure as the end pocket in the caecum of herbivorous mammals. Its presence in humans suggests a function in early human evolution and is evidence that humans share a common ancestry with herbivorous mammals.
- **Analogous features**: perform the same function but are fundamentally different in structure, for example, the wings of butterflies and birds. Analogous structures suggest that the organisms are not closely related. Such features are the result of convergent evolution.

MOLECULAR HOMOLOGY

Molecular homology is the similarity in the molecular characters (e.g. DNA and proteins) of organisms due to common ancestry. Molecular homology provides evidence of evolutionary relationships between species.

- A genetic linkage group is a group of gene loci that are in close proximity to one another on a given chromosome. Examination of similar linkage groups (genes controlling the same characteristics) in different mammals reveals the same pattern of linkage.
- Amino acid sequences: comparison of the order of amino acids in polypeptide chains can be used to determine evolutionary relationships. Some closely related species have identical amino acid sequences in functionally important proteins, for example cytochrome c and haemoglobin in humans and chimpanzees.
- DNA sequences: comparison* of the order of nucleotides is a useful tool in gauging the degree of similarity between species and the closeness of their evolutionary relationships (Table 4.2.4). Various sources are available for DNA sequencing—nuclear DNA, mitochondrial DNA and chloroplast DNA.
    - Nuclear DNA is used to sequence entire genomes.
    - Mitochondrial DNA (mtDNA) undergoes change (mutation) rapidly compared to nuclear DNA. Comparing mtDNA of related species is a useful tool in estimating the timeframe for the divergence of species that have separated in evolutionary terms over a relatively short time (about 20 million years).
    - Chloroplast DNA in plants and algae is a short, circular molecule that undergoes evolutionary change relatively slowly, making it useful in estimating when divergence occurred between groups over a large period of geological time, for example flowering plants (angiospermae) and conifers (gymnospermae).

In summary, the conservation of similar genes in different species, and of polypeptides that are the expression of those genes, is strong evidence that species are related in evolutionary terms and share common ancestors.

Table 4.2.4 Comparison of DNA of different species using DNA sequencing

% similarity between human genome and	
chimpanzee	98
gibbon	96
capuchin monkey	84

KEY KNOWLEDGE

PATTERNS OF EVOLUTION

- **Divergent evolution**: species that share a common ancestral species have diverged over time, evolving into distinctly different species, for example, the body shape and flippers in dolphins and seals (Figure 4.2.5).

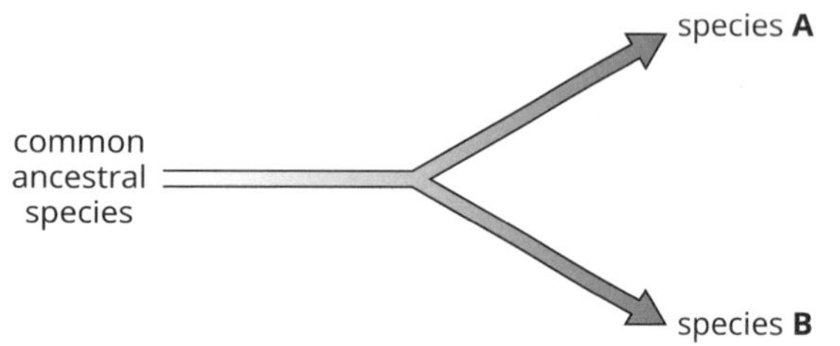

Figure 4.2.5 Divergent evolution

- **Convergent evolution**: unrelated species display similar features as a result of similar environmental pressures, for example, the body shape of dolphins and sharks (Figure 4.2.6).

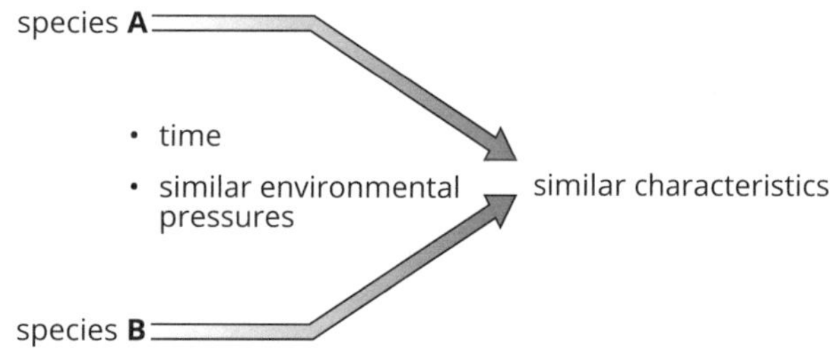

Figure 4.2.6 Convergent evolution

- Parallel evolution: related species that share a common evolutionary ancestor are subsequently subject to similar environmental pressures that lead to similar features (Figure 4.2.7). An example is the evolution of the patagium (a flap of skin between the forelimbs and hindlimbs used to 'glide') present in the Australian sugar glider (marsupial mammal) and the southern flying squirrel of North America (placental mammal).

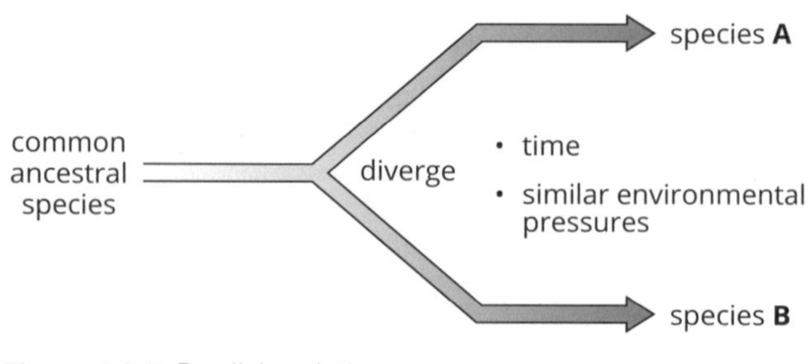

Figure 4.2.7 Parallel evolution

MOLECULAR CLOCK

A **molecular clock** shows the rate of mutational change observed in the different sources of DNA molecules of species over time. It can be used to measure the degree of genetic similarity and difference between species, and to estimate the time at which related species diverged. This provides a picture of the degree of evolutionary relationships between species. Mitochondrial DNA and chloroplast DNA are examples.

PHYLOGENETIC TREES

Phylogeny is the pattern of evolutionary relationships. Evidence from structural morphology and molecular characters helps scientists to build a picture of the evolutionary relationships between different species of organisms, and between the different groups of species. Such relationships can be represented in diagrams called **phylogenetic trees** (Figures 4.2.8 and 4.2.9).

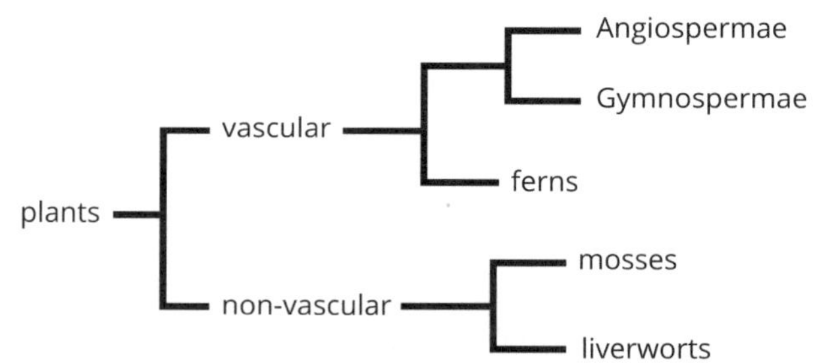

Figure 4.2.8 Phylogenetic tree of plants

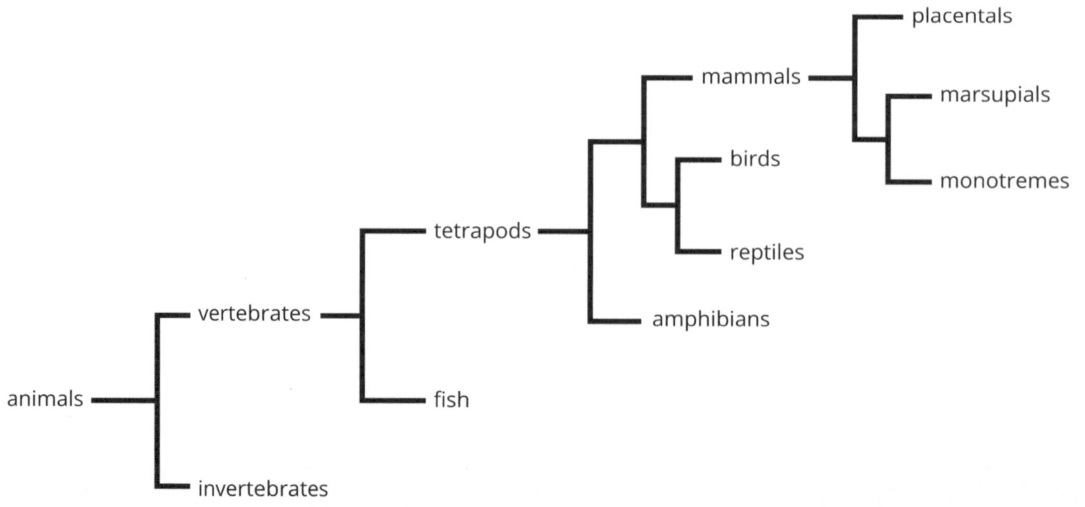

Figure 4.2.9 Phylogenetic tree of vertebrates

 ISBN 978 0 6557 0026 5

KEY KNOWLEDGE

Human change over time

Our own species, *Homo sapiens*—modern humans, has existed on Earth for about 200 000 years.

Humans belong to various taxonomic groups and share the characteristics of these groups.

- Class mammalia: characterised by features such as mammary glands to produce milk for infants, a four-chambered heart, different kinds of teeth and a covering of hair or fur (land mammals). Virtually all mammals are endothermic.
- Order Primates: characterised by features such as a grasping hand, well-developed brain, flattened face, binocular vision, short nose and bicuspid teeth. Humans share the order Primates with lorises, lemurs, monkeys, gibbons and the great apes, for example, chimpanzees, gorillas and orangutans.
- Family **Hominidae**: characterised by features such as a well-developed brain with complex cerebral cortex, characteristic upper jaw and teeth and characteristic frontal bones in the skull. Humans share the family Hominidae with two species of chimpanzee, gorillas and orangutans.

There is only one surviving species in the genus *Homo*—*H. sapiens*. It is important to distinguish between hominoids (primate superfamily), **hominids** (includes humans and the great apes) and **hominins** (includes only modern and extinct humans). The distinguishing features of *H. sapiens* include:

- a fully upright stance (bipedal)
- fewer and smaller teeth than other members of family Hominidae
- a relatively flattened facial profile
- reduced brow ridges
- a relatively large braincase (1200–1500 cm^3)
- well-developed cultural features including the ability to make tools, use complex language and other forms of communication such as writing, art and rituals.

HUMAN EVOLUTIONARY HISTORY

In the relatively recent geological past, several different species of the genus *Homo* coexisted at various times with *H. sapiens*. These include *Homo neanderthalensis*, *Homo denisova* and *Homo floresiensis*.

As well as this, over the last 5 million years, various species of hominins existed. Some of these represent ancestral species from which anatomically modern humans evolved (Table 4.2.5). Others diverged from a common ancestral species.

Table 4.2.5 Ancestral species

Hominin	Features
Sahelanthropus	• lived 6–7 million years ago in Africa • ape-like braincase • human-like facial structure and teeth • features that indicate bipedalism and tree-climbing
Ardipithecus	• lived 4.4–5.8 million years ago in Africa • features that indicate bipedalism and tree-climbing • body and brain size similar to chimpanzees • canines similar to those of later hominins
Australopithecines	• various species lived between 2.5 and 4.2 million years ago • 'robust' hominins, heavy boned with pronounced brow ridges • smaller braincase than modern humans • teeth smaller than the apes • fully bipedal with arms relatively long, but not longer than legs
Paranthropus	• three species that lived between 1.6 and 2.8 million years ago • 'robust' hominins • fully bipedal • used tools
Homo	• at least eight recognised species that lived between 2 million years ago and the present day • relatively larger braincase than predecessors, general increase in braincase size with subsequent *Homo* species • fully bipedal with arms shorter than legs • 'gracile' hominins, less heavy boned than 'robust' counterparts • only one surviving species—*H. sapiens*

KEY KNOWLEDGE

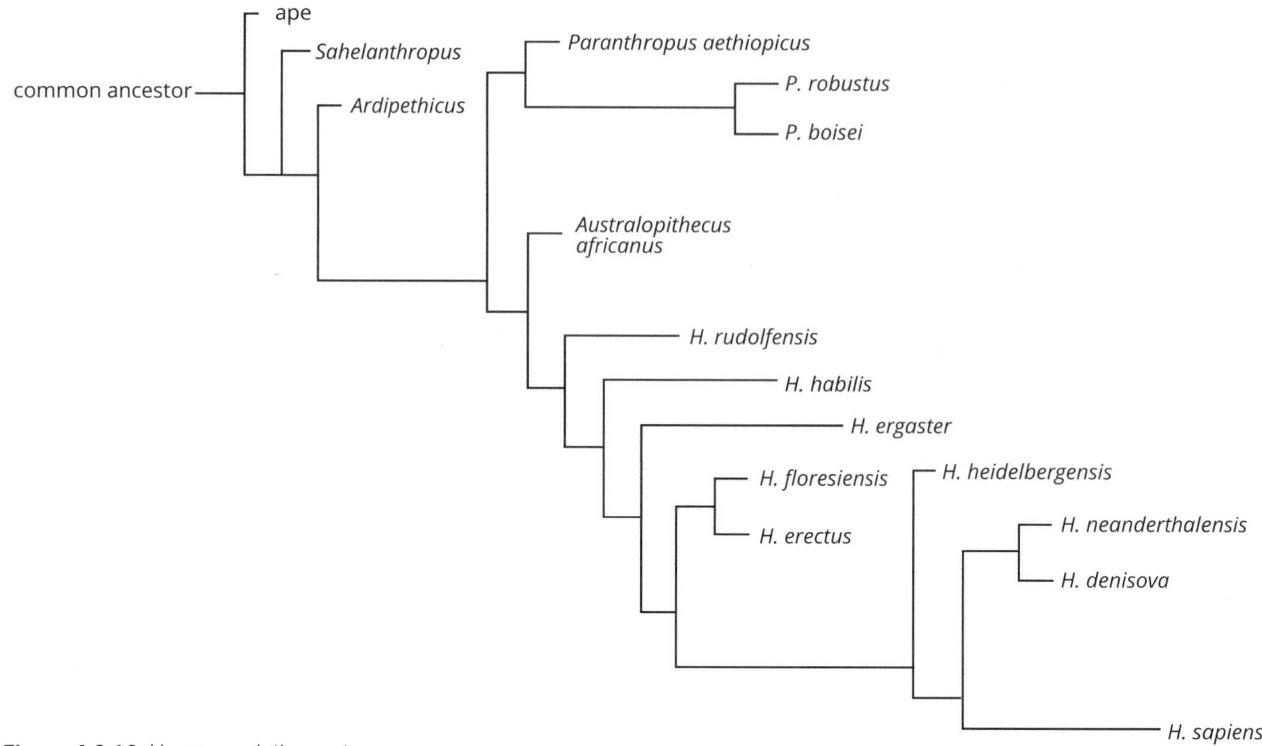

Figure 4.2.10 Human evolutionary tree

Table 4.2.6 Comparison of species in genus *Homo*	
Species	**Distinguishing features**
Earliest species had a relatively small braincase; did not use tools	
H. rudolfensis	Africa
H. habilis	Africa
H. ergaster	Africa
H. erectus	Asia; walked upright
More recent species	
H. heidelbergensis (Germany)	• used tools and fire • hunted large animals • built shelters
H. neanderthalensis (Germany)	• 200 000 years ago until 30 000 years ago • braincase slightly larger than modern humans • pronounced brow ridges • used tools and fire • culture included ritual burial • language
H. denisova (Siberia)	• 40 000 years ago • little known species (fossils limited to a single finger bone and two teeth)
H. floresiensis (Indonesia)	• as recently as 18 000 years ago • small braincase: 380 cm^3 • used tools
H. sapiens	• see page 153 for distinguishing features

These hominins are represented by different genera as shown in Figure 4.2.10.

Table 4.2.6 lists some of the distinguishing features of the genus *Homo*.

Different theories have been proposed to account for the evolution of modern humans. There are three widely accepted theories.

- The **Multiregional evolution (continuity) model** states that *Homo erectus* migrated across Africa, Asia and Europe during the last 1.8 million years. Isolation of the populations resulted in the divergence of gene pools but occasional contact ensured some gene flow was maintained and led to the concurrent evolution of all groups as one recognisable species—*H. sapiens*.
- The **Out of Africa (replacement) model** states that populations of modern humans originated from a discrete population in Africa about 200 000 years ago, and then migrated to other continents in more recent times.
- The **Assimilation (partial replacement) model** states that modern humans had an African origin and when people migrated out of Africa there was occasional interbreeding with archaic humans who were already living in other parts of the world, resulting in hybrid populations (assimilation).

These three dominant theories of modern human origins all support an African origin but have different views on the pattern of migration and interbreeding that occurred in other parts of the world.

 ISBN 978 0 6557 0026 5

KEY KNOWLEDGE

Our understanding of the evolution of our own species is also subject to interpretation and refinement as new fossil evidence is uncovered. Paleoanthropologists analyse the evidence to construct the most current picture possible to explain our human origins and the evolutionary pathways of the various human species. Interactions between members of different *Homo* species is also the subject of speculation, with some evidence suggesting interbreeding between *H. sapiens* and *H. neanderthalensis*.

HUMAN MIGRATION TO AUSTRALIA

Comparative genomic data from indigenous populations from Australia and Papua New Guinea, Asia, Europe and Africa reveal that ancestors of Indigenous Australians migrated from Africa around 70 000–75 000 years ago, reaching the Australian/Papuan landmass (Sahul) around 50 000 years ago, when sea levels were lower and the lands were connected. By 35 000–40 000 years ago, Indigenous Australian populations were geographically separated from Papua New Guinea and migrating across the Australian continent. Genomic evidence indicates genetic diversity already existed between Indigenous Australian populations as long as 31 000 years ago. Before European settlement, Indigenous Australians were represented by more than 250 nations, each distinguishable by language and culture as well as unique custodianship of Country.

Researchers investigating the genetic and geographic structure of Indigenous Australian populations have found evidence for a strong connection between populations and discrete geographic areas, supporting the important cultural connection to Country and Place still held by Indigenous Australians today.

HUMAN CULTURE AND INTERVENTIONS

A key feature of modern humans is the development of complex culture. Human populations have developed permanent settlements using agriculture to provide food as an alternative to the hunter–gatherer lifestyle, and have manipulated the global environment and resources to enhance their lifestyle (Table 4.2.7).

The earliest evidence of the emergence of modern behaviour and culture in *Homo* is from tools found in Africa dating to approximately 300 000–400 000 years ago. Blades, bone tools, specialised weapons for hunting and evidence of long-distance travel and trade signify the beginnings of cultural development in humans. The sophisticated use of tools was crucial to the evolution of human culture for hunting, adornment, ritual, clothes making, agriculture, writing, building and many other important advances. The use of tools marked the beginning of cultural evolution in *Homo*.

Evidence of the use of ochre (clay in shades of yellow, brown, red and purple) and pierced shells for adornment date back to 100 000–200 000 years ago in sub-Saharan Africa. Such artefacts indicate that culture, in the form of ritual and symbolic expression, may have begun to emerge much earlier than the cultural explosion that was thought to have occurred 50 000 years ago.

More complex forms of symbolic expression are evident in cave paintings and carvings that became prolific about 35 000–40 000 years ago. Paintings in a cave in Sulawesi, an island in Indonesia, date back to 39 900 years ago, making them the earliest known forms of artwork.

Table 4.2.7 Human culture and interventions

Feature	Examples
communication	• complex forms of communication, including language, writing and art • information is passed from one generation to the next
agriculture	• allows manipulation of environment and resources that leads to more sedentary lifestyle with more time to develop other pursuits (e.g. art, communication)
selective breeding	• breeding programs designed to breed plants and livestock with the most desirable characteristics • alters allele frequencies in populations; this may impact on wild populations by limiting the gene pool
science and technology	• includes advances in various forms: – electronic communications/digital age – development of pesticides and herbicides to enhance agricultural output – medical interventions aimed at treating and preventing disease (e.g. antibiotics)
intervention in evolutionary processes	• the development and application of gene technologies, such as bacterial transformation to manufacture insulin • use of stem cell technology in disease research • genetic screening in determining carrier status for particular inherited disorders • gene therapy in treating genetic diseases • DNA profiling

WORKSHEET 28

Knowledge review—evolution foundations

1 The terms below will be familiar to you from themes you have studied before. Explain what is meant by each term so that a younger student will understand them. Discuss your responses with other students. Consult resources to check the accuracy of your answers. Finally, refine your explanation.

a species

b fossil

c mutation

d evolution

e natural selection

f biological fitness

2 Humans belong to the family Hominidae, which also includes chimpanzees, gorillas and orangutans. Think about the ways in which we are similar to these great apes and what makes us different from them. Summarise your understandings in the table. When you have completed your responses, share them with other students. Refine your responses.

In relation to the great apes, describe how humans are:	
similar	
different	

ISBN 978 0 6557 0026 5

WORKSHEET 29

Evolutionary change over time

Ecosystems are dynamic in nature, with change occurring along a continuum. Small-scale changes that occur over very short periods of time tend to be less obvious than large-scale change that has occurred over long periods. Close examination of ecosystems reveals a rich story of change. The task of the scientist is to know where to look and what questions to investigate.

Examine the images below, then answer the questions about what they reveal about past change in the ecosystems they represent.

1 The Aboriginal rock art shown in Figure 4.2.11 is at Ubirr in Kakadu National Park in the Northern Territory. It has been radiometrically dated to between 1500 and 28000 years old.

a Suggest the kind of animal this rock art depicts. Give reasons for your answer.

Figure 4.2.11 Aboriginal rock art in Ubirr, Kakadu National Park, Northern Territory

b What does this art suggest about change in the distribution of this species over time?

c How might we account for this change?

2 Radiometric dating of this large predator fish, the Cooyoo, indicates that it lived over 100 million years ago (Figure 4.2.12). Its fossilised remains, along with the fossils of many other fish and other marine organisms, have been uncovered across a vast area of inland northwest Queensland.

a What does this evidence suggest about ecosystem change in this region over the last 100 million years?

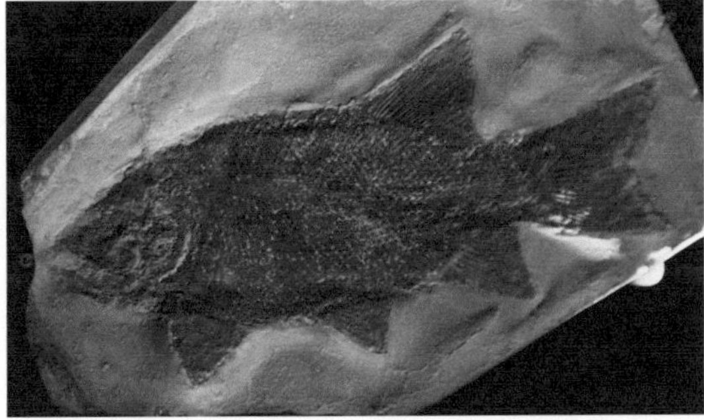

Figure 4.2.12 Fossilised remains of a Cooyoo fish

b Describe a change in an abiotic factor that may have led to ecosystem change and extinction for this species.

3 Analysis of air bubbles trapped in Antarctic ice and extracted in ice core samples is represented in Figure 4.2.13.

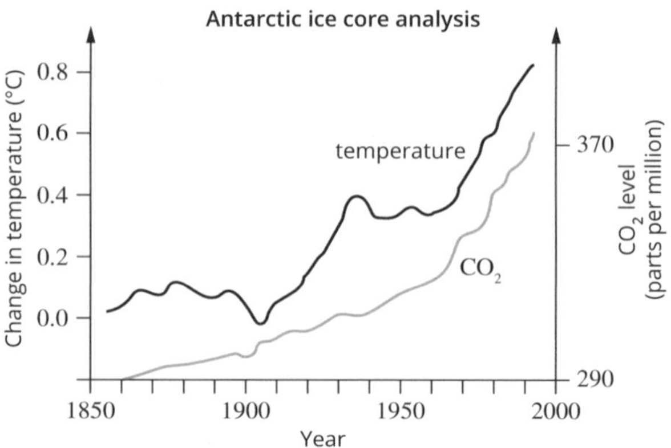

Figure 4.2.13 Analysis showing the relationship between temperature change over time and the level of CO_2 present in Antarctic ice core samples

a Describe the changes in atmospheric temperature between the years 1850 and 2000.

b Describe the changes in atmospheric carbon dioxide (CO_2) concentration in this time.

c Outline the relationship the graph suggests between atmospheric carbon dioxide concentration and temperature.

4 The photograph in Figure 4.2.14 illustrates coral bleaching of an area of the Great Barrier Reef. When coral polyps are stressed, the algae they contain are expelled and the coral becomes white or bleached. Without the algae, the life of the coral is threatened. A significant decline in corals has been measured between 1985 and 2010. Scientists believe this is largely due to cyclones, ocean acidification, warmer waters due to global warming, and the crown-of-thorns sea star.

Suggest how the environmental change of rising carbon dioxide concentrations contributes to change in the coral population.

Figure 4.2.14 Bleached coral on the Great Barrier Reef, Queensland

 ISBN 978 0 6557 0026 5

WORKSHEET 30

Changing allele frequencies in a population

The phenotypes in populations change over generations and time. This is related to changes in the allele frequencies from generation to generation. Changes in allele frequencies are the result of various factors including environmental selection pressures, chance environmental events, changes in chromosome number and mutations.

1 Consider each scenario below. Select from the list to accurately identify the process that has occurred in each instance to account for the change in allele frequencies in the population.

mutation	artificial selection	genetic bottleneck	founder effect	environmental pressure

	A small population of Dutch colonists, called Afrikaners, settled in South Africa in 1652. One of the men carried the allele for Huntington's disease—a rare neurodegenerative condition that affects individuals later in life, typically after the reproductive years. Huntington's disease is an autosomal dominant disorder. The allele was passed on to subsequent generations. Huntington's disease is now over-represented in South Africa's Afrikaner population compared to the Dutch population in Holland.	
	A cat breeder with a preference for black coat colour chooses only black-coated individuals for breeding.	
	A change in the DNA of pathogenic *Staphylococcus* bacteria results in resistance to antibiotic drugs.	ANTIBIOTIC Antibiotic resistant bacteria
	With estimates of less than 20 individuals remaining in the wild, the orange-bellied parrot is a Critically Endangered Australian species. With such small numbers, the genetic diversity of the population has declined.	
	In 1996 the rabbit calicivirus was released in Australia to control soaring rabbit populations. Most rabbits were sensitive to the disease and died, markedly reducing rabbit populations.	

2 **a** Explain how the founder effect described in the example above is likely to contribute to a reduced genetic diversity for this population.

______

______

b Outline the potential consequences for this population as a result of reduced genetic diversity.

______

______

WORKSHEET 31

Variation, selection and other factors driving change in populations

1 Use a ruler and pencil to match each of the statements on the left with its corresponding example on the right.

Genetic diversity exists within populations.	Two pink-flowering snapdragon plants produce some red-flowering and some white-flowering offspring.
Members of species show **adaptations** to their environments.	A dairy farmer deliberately retains cows for breeding that produce large volumes of milk.
Sexual reproduction results in offspring that are different from their parents and from one another.	Dieldrin-resistant blowflies increase in a population exposed to the pesticide, while the frequency of sensitive individuals declines.
Natural selection results in an increase in the frequency of **biologically fit** individuals, i.e. those with features that are suited to the environment, compared to individuals that do not have these features.	A small population that is isolated on an island shows change in allele frequency resulting from chance only.
Human intervention in the form of **artificial selection** can alter the genotype and phenotype of populations of organisms.	Migration from one population into another introduces new alleles into the second population.
Changes in populations can occur as a result of **genetic drift** or the **founder effect**.	Feather colouration in the tawny frogmouth makes it well camouflaged in its natural environment.
Dispersal of seed and pollen by water, wind and animals leads to **gene flow** between populations of plants.	A population of rabbits has some individuals that are resistant to calicivirus and some that are sensitive.

2 Provide definitions for each of the terms that appears in bold type.

ISBN 978 0 6557 0026 5

Modelling

Alterations in wild gene pools by selective breeding

Unlike natural selection, artificial selection involves the human manipulation of an organism's gene pool with the intention of breeding organisms with the most desirable characteristics to accommodate a particular human need or purpose. Selective breeding programs mean only those individuals displaying the most desirable characteristics are allowed to breed. Humans have selectively bred plants and animals for thousands of years, a process that has led to the domestication of many wild species.

The domestication of cattle began some 10 500 years ago in Southwest Asia. The ancestral species from which today's cattle are descended is an extinct species of large wild cattle called the aurochs. Selective breeding of cattle with particular desirable features has resulted in the wide variety of different cattle breeds we have today. Notable examples include Friesian cows renowned for their high milk yield and the muscular form of Belgian Blue beef cattle for eating (Figure 4.2.15).

Figure 4.2.15 The domestication of Friesians and Belgian Blue beef cattle from an ancestral species of wild cattle

1 Prepare a flow chart to summarise the steps in selective breeding that might have lead to the production of high milk-yielding cattle from an original herd. A start has been made for you.

Variation exists in the herd in relation to milk yield—some cows produce higher volumes of milk than others.

2 Comment on the degree of variation likely to exist within each of the domesticated cattle breeds compared with the wild cattle from which they have descended.

3 Outline the potential consequences for each breed in the event of environmental challenges such as disease.

WORKSHEET 33

Patterns in evolution

1 Evaluate the accuracy of each of the statements below. Circle either TRUE or FALSE for each statement.

- Evolution involves change in populations of organisms over time and generations. TRUE/FALSE
- Divergent evolution occurs when related species diverge from a common ancestor and then independently evolve similar features. TRUE/FALSE
- Convergent evolution occurs when unrelated species display similar adaptations. The presence of similar adaptations is attributed to different groups being subjected to similar environmental pressures. TRUE/FALSE
- The presence of 13 different finch species on the Galápagos Islands is understood to be an example of convergent evolution. TRUE/FALSE
- Parallel evolution occurs when different but related species diverge from a common ancestral group. TRUE/FALSE
- The rapid divergence of one or more species from a common ancestral group is referred to as adaptive radiation. The different finch species on the Galápagos Islands are an example of adaptive radiation. TRUE/FALSE
- Populations of the same species living in different geographic locations and subject to different environmental pressures are likely to show a variation in the frequency of the same alleles. TRUE/FALSE
- When populations of the same species are geographically isolated over long periods of time and subject to different environmental pressures, the populations become increasingly adapted to their respective environments and increasingly different from each other. If members of the different groups become so different that reproduction is no longer possible between members of the different groups, allopatric speciation is said to have occurred. TRUE/FALSE
- Mechanisms that maintain reproductive isolation between related species include behavioural mechanisms such as differences in mating calls and structural differences in genitalia that physically rule out reproduction. TRUE/FALSE
- The evolution of two species from an ancestral one in the same geographic area is called sympatric speciation. TRUE/FALSE

 ISBN 978 0 6557 0026 5

2 Examine each FALSE statement. Rewrite each as a TRUE statement.

ISBN 978 0 6557 0026 5

WORKSHEET 34

Evolution essentials

Complete the crossword puzzle to help check your knowledge and understanding of key terms and processes related to evolution.

Across

5 Preserved remains, impressions, casts or other evidence of the past existence of organisms [6]

6 Dating technique that analyses light emissions from minerals contained in human artefacts [18]

7 Precambrian cyanobacteria that form thin mats, trapping sediment and thereby producing a concentric arrangement of layered rock [13]

8 Fossil of known age used as a point of comparison to determine the approximate age of rock strata in different locations [9, 6]

9 Dating technique that analyses the half-life of a particular element—limited in its dating capacity to about 60 000 years before present [11, 6]

10 Time that elapses for half of the atoms in an isotope sample to decay [4, 4]

11 Comparative study of rock strata and the fossils they contain to determine the relative age of fossils and rock [12]

15 Three-dimensional structure taking the form of an organism that once formed an impression in rock [4]

Down

1 Process of change in organisms over time [9]

2 Dating techniques that provide accurate assessments of the age of fossils are referred to as ______________ dating techniques [8]

3 Describes patterns of evolutionary relationships; sometimes represented using branching tree diagrams [9]

4 Dating technique that uses calculation of half-life of an isotope to determine the precise age of a fossil [11, 6]

9 Comparative method of dating fossils, for example, in relation to previously dated fossil or rock stratum in the vicinity of unknown fossil [8]

12 Extinct early Cambrian marine arthropod [9]

13 Radioactive form of an element [7]

14 Tree resin or sap that sometimes traps small organisms such as insects, which are subsequently fossilised in this form [5]

ISBN 978 0 6557 0026 5

WORKSHEET 35

The Galápagos finches

The 13 species of Galápagos finches represent an example of evolutionary radiation. Populations from the ancestral species that were subjected to different environmental pressures on the different islands, and different local environmental pressures on the same islands, eventually resulted in the different species that exist today.

Use your knowledge and understanding of natural selection and speciation to summarise the key steps in the possible evolutionary pathway that led to the presence of different species of finches from an ancestral population on the Galápagos Islands (Figure 4.2.16).

Include the terms listed below in your response.

microevolutionary change	mutation
genetic divergence	isolation
natural selection	allopatric speciation
gene flow	variation
macroevolutionary change	

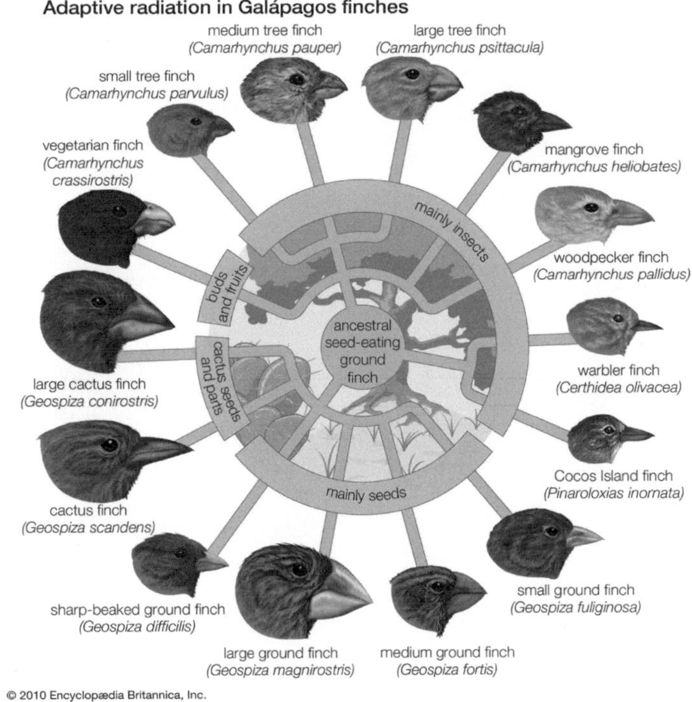

Figure 4.2.16 Adaptive radiation of Darwin's Galápagos finches

WORKSHEET 36

Evidence of evolution from morphology and molecules

1 Read the definitions listed on the right side of the page. Choose the correct term from the box below and write it beside its corresponding definition.

molecular clock structural morphology chloroplast DNA analogous structures
tetrapod phylogenetic tree genetic linkage group vestigial structure fossil DNA
homologous structures molecular homology mitochondrial DNA

	DNA contained within the organelles that carry out cellular respiration
	name of metaphorical timepiece that describes the measure of mutational change over time in DNA of related species to estimate time since divergence of groups
	similar anatomical features of unrelated organisms that serve the same function
	group of gene loci positioned closely together on a particular chromosome
	anatomical features with fundamental structural similarity between related organisms
	DNA contained within the organelles that carry out photosynthesis
	study of anatomical features of different organisms to identify similarities and differences in structure to determine evolutionary relationships
	comparison of organisms' DNA or proteins to identify degree of similarity and thereby establish the closeness of their evolutionary relationship

2 You will see that there are four terms left over. Write a definition for each of these terms.

 ISBN 978 0 6557 0026 5

Comparing structural morphology

Mammals, birds, reptiles and amphibians are very different groups of vertebrates. Even within each group there is an enormous diversity of forms, yet the similarities are conspicuous.

1 The diagrams below represent the forelimbs of seven different vertebrates belonging to these groups.

a Compare the forearm anatomy of the different organisms to help you identify and label the bones that have not been labelled.

b Use coloured pencils to code the bones. For example, colour all the humerus bones in red.

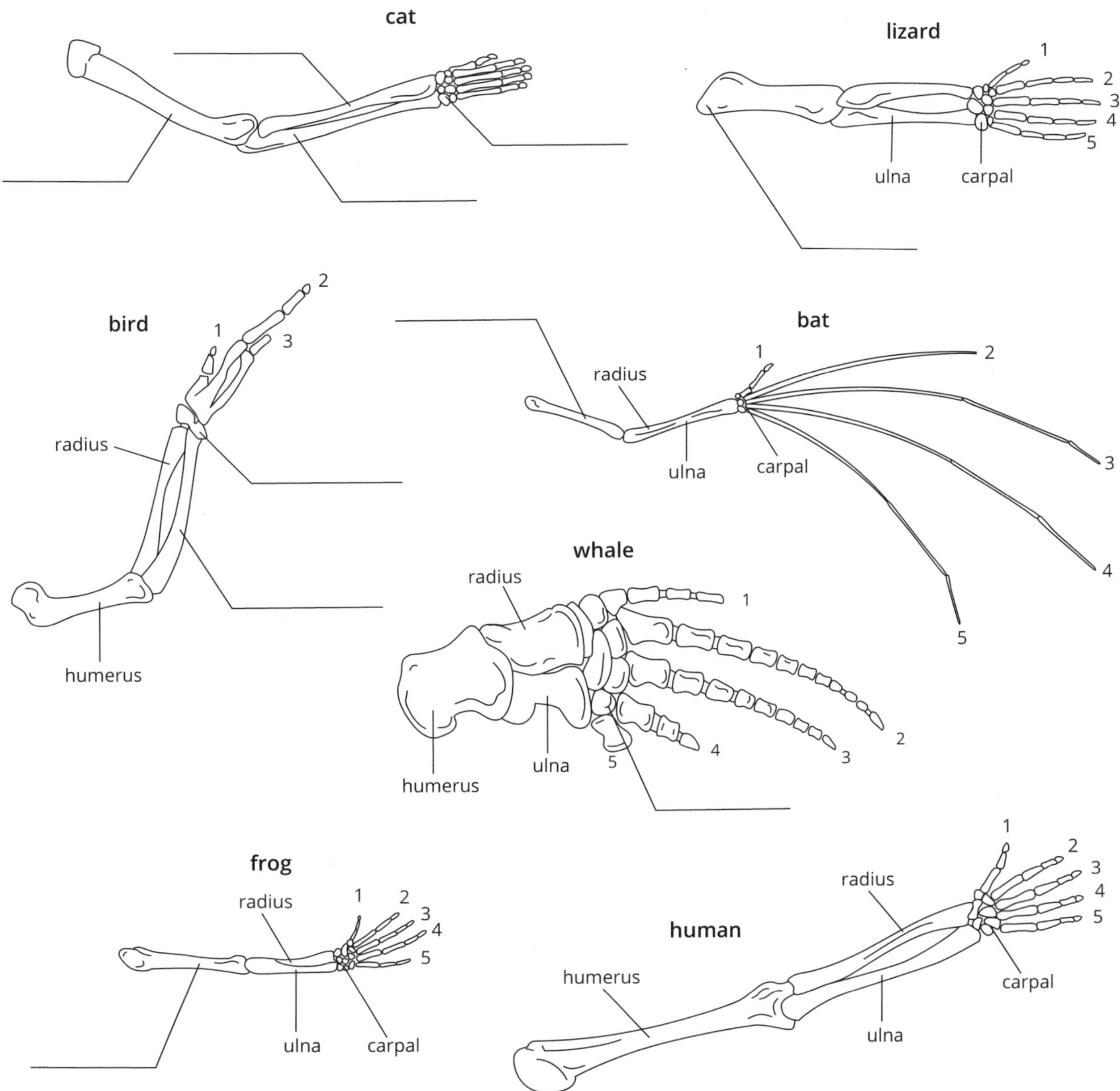

2 The forelimbs of these tetrapod vertebrates are called homologous structures. Explain what is meant by this term.

3 Outline what these homologous structures suggest about the evolution of these different organisms.

 ISBN 978 0 6557 0026 5

WORKSHEET 38

Human evolution

Use your knowledge and understanding of human evolution to write your own clues for each of the terms presented in the crossword.

Across

4 ______

5 ______

6 ______

7 ______

8 ______

9 ______

10 ______

Down

1 ______

2 ______

3 ______

11 ______

WORKSHEET 39

Reflection—How are species related over time?

The following table lists the key knowledge covered in this area of study.

1 Reflect on how well you understand the concepts listed. Rate your learning by shading the circle that corresponds to your current level of understanding for each one.

Key knowledge	Not confident ◄			► Very confident
Causes and consequences of changing allele frequencies	○	○	○	○
Selective breeding	○	○	○	○
Bacterial resistance and viral antigenic drift and shift	○	○	○	○
Evidence of changes in species over time from the fossil record	○	○	○	○
Speciation as a consequence of isolation and genetic divergence	○	○	○	○
Evidence of the relatedness between species—structural morphology and molecular homology	○	○	○	○
Phylogenetic trees	○	○	○	○
Shared characteristics that define mammals, primates, hominoids and hominins	○	○	○	○
Evidence for the major trends in hominin evolution	○	○	○	○
The human fossil record as a classification scheme that can be contested and refined	○	○	○	○
Fossil and DNA evidence to explain the migration of modern humans	○	○	○	○

2 Consider the points you have shaded from Not confident to Very confident. List specific ideas you can identify that were challenging.

3 Write down two different strategies that you will apply to help further your understanding of these ideas.

ISBN 978 0 6557 0026 5

PRACTICAL ACTIVITY 12

Case study • Modelling

Natural selection in an island population

Suggested duration: 60 minutes

AIM

- To investigate the relationships between the incidence of an inherited blood disorder (thalassaemia) and the non-inherited disease, malaria.
- To consider the selection forces acting to maintain normally disadvantageous alleles at relatively high frequencies in populations.

Thalassaemia

The oxygen-carrying pigment in red blood cells is called haemoglobin. A normal haemoglobin molecule is made up of four small protein chains. There are two alpha (α) chains and two beta (β) chains that fit together into a specific three-dimensional shape that optimises haemoglobin's oxygen-carrying capacity.

In some individuals, the beta haemoglobin molecules are abnormal, resulting in a reduced oxygen-carrying capacity. This condition is called thalassaemia. There are at least two allelic forms of the gene for beta haemoglobin—$\beta+$ is the allele for normal haemoglobin production, and $\beta-$ is a mutant allele that reduces beta haemoglobin production. The gene for beta haemoglobin production is located on an autosome. If only one of the normal alleles is present, reduced levels of beta haemoglobin are produced, resulting in distorted but not seriously compromised red blood cells that function fairly normally. When individuals carry this $\beta-$ allele, they suffer from mild anaemia.

Prior to 1976, no treatment was available for sufferers of thalassaemia major and death before eight years of age was certain. Today, modern technologies provide treatments that can prevent serious disability and even death.

1 Using the notation suggested, write down all the possible genotypes that can be produced from the different combinations of the two alleles of this gene.

2 Individuals who are homozygous for the mutant allele have a condition called thalassaemia major. Suggest how such individuals are likely to be affected.

3 Suggest an appropriate name for the less serious form of thalassaemia, which occurs in the heterozygous condition.

4 Name a management strategy available through modern technology and outline how it improves the health and life expectancy of sufferers of thalassaemia major.

Malaria and thalassaemia—an unlikely union

Malaria is a non-inherited disease caused by the parasite *Plasmodium falciparum*, which spends part of its life cycle in red blood cells. The transmission agent for malaria is the *Anopheles* mosquito, which must have access to water for much of its life cycle.

Thalassaemia and another blood cell disorder, sickle cell anaemia, are common disorders in Africa, the Mediterranean and parts of Asia. Before 1946, they were widespread (Figure 4.2.17). Interestingly, these are also areas in which malaria was prevalent at the time.

Several detailed studies have been made of the incidence of malaria and thalassaemia on the Mediterranean island of Sardinia, located off the coast of Italy. The island has swampy coastal plains and a high central mountain range. Many small villages are effectively isolated from one another.

The results of studies considering the incidence of thalassaemia in a number of small villages roughly along a transect line from Oristano to Posada (Figure 4.2.18) are shown in the graph in Figure 4.2.19.

Thalassaemia is a genetic blood disorder that adversely affects an individual's chance of survival, so we would expect the frequency of the alleles responsible for this and other similar conditions to decline within populations. However, the data suggests that this is not the case in Sardinia.

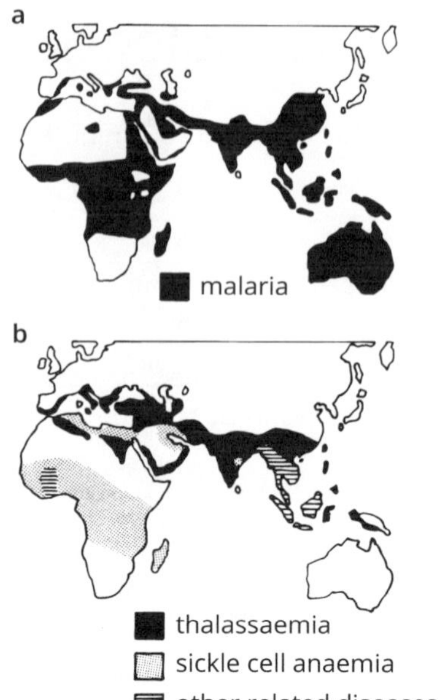

Figure 4.2.17 Comparison of (a) the distribution of malaria with (b) the incidence of inherited blood disorders

5 Suggest a logical explanation to account for the reduction in the incidence of malaria in the regions listed since 1946.

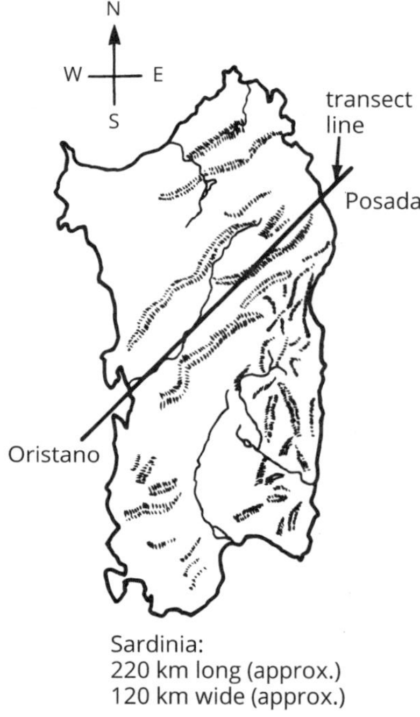

Figure 4.2.18 Map of Sardinia showing mountain ranges and approximate position of the transect line

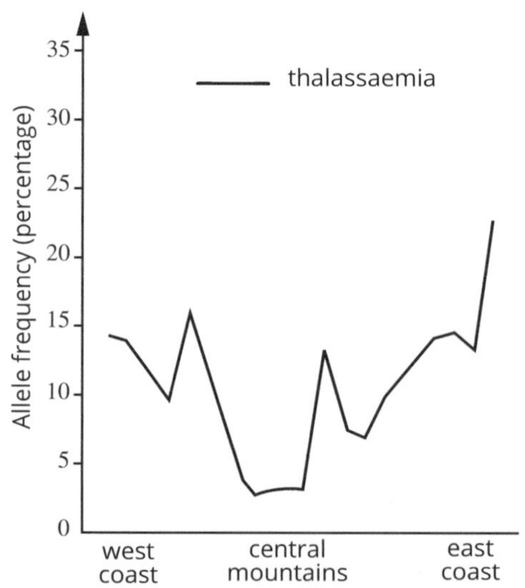

Figure 4.2.19 Frequency of thalassaemia along a transect line across Sardinia

 ISBN 978 0 6557 0026 5

PRACTICAL ACTIVITY 12

6 Graph the frequency of malaria in relation to altitude using the information in the table below. Label the axes on your graph.

Percentage frequency of malaria in a number of Sardinian villages situated at different altitudes		
Village	**Altitude (m)**	**Malarial frequency (%)**
Benetutti	406	50
Cabras	9	97
Desulo	891	33
Fonni	1000	19
Gatellì	40	96
Gergei	374	60
Isii	523	17
Lanusei	595	25
Lodè	345	90
Orosei	19	93
Suni	333	85
Teulada	50	90

7 Describe the relationship between the incidence of malaria and the altitude of populations in Sardinia.

8 Look carefully at the data in the graph in Figure 4.2.19 and in the graph you have drawn. Write a sentence comparing the two sets of data.

9 Suggest how the presence of abnormal red blood cells affects an individual's resistance to malaria. (Clue: Find out about the life cycle of the malaria *Plasmodium*.)

10 Explain whether the shift in allele frequencies for thalassaemia between populations at different altitudes on the island of Sardinia (observed in phenotype differences) represents an example of microevolution or macroevolution.

CONCLUSION

11 Describe how an environmental pressure—in this instance, the presence of the malarial parasite—has led to an increase in the incidence of an otherwise disadvantageous inherited condition, thalassaemia. Include a discussion of:

- variation
- allele frequencies
- phenotypes that confer a survival advantage given certain environmental pressures
- change in the phenotypic ratio of a population over time.

 ISBN 978 0 6557 0026 5

PRACTICAL ACTIVITY 13

Divergent and convergent evolution in anteater species

Suggested duration: 60 minutes

INTRODUCTION

Have you ever wondered why some apparently unrelated animals show quite remarkable physical likenesses? A bat and a bird, a dolphin and a shark, a sugar glider and a flying squirrel—each pair shares some very obvious structural characteristics. And yet within any one group of animals there can be a remarkable diversity of forms. The most famous example of this was first described by Charles Darwin when he visited the Galápagos Islands in the 1800s. He recorded an astonishing array of beak shapes in the finches native to those islands.

Two evolutionary patterns are operating here. The first is convergence—the evolution of similar structures in organisms that do not share a recent common ancestor. The second is divergence—the evolution of different structures in organisms that share a recent common ancestor.

Examine the body outlines and read the information related to each of the different anteater species. Anteaters are mammals that feed exclusively on ants and/or termites.

Echidna (*Tachyglossus*)

The Australian echidna (Figure 4.2.20) is a monotreme (egg-laying) mammal. It has an elongated, beak-like muzzle, with a small mouth at the end. It lacks teeth, but has a long, sticky tongue. It has small eyes and no external ears. The short limbs are equipped with strong claws. The upper surface of its body is covered with heavy spines and it lays leathery, shelled eggs.

Head and body length: 25–53 cm

Body weight: 2.5–6.0 kg

Figure 4.2.20 *Tachyglossus* (Australian echidna)

Aardvark (*Orycteropus*)

The African aardvark (Figure 4.2.21) is a placental mammal. It has a slender skull with no canine or incisor teeth. Its sticky tongue is extensible and elongated. Its short limbs end in digits with strong claws. It is a lightly haired animal with long ears and small eyes.

Head and body length: 100–158 cm

Body weight: 50–70 kg

Figure 4.2.21 *Orycteropus* (African aardvark)

Giant anteater (*Myrmecophaga*)

The South American giant anteater (Figure 4.2.22) is a placental mammal. Its skull has an elongated snout and there are no teeth present. The tongue is long and sticky. Its body is covered with long hair. The forelimbs are each equipped with four strong claws.

Head and body length: 100–120 cm

Body weight: 18–23 kg

Figure 4.2.22 *Myrmecophaga* (South American giant anteater)

Pangolin (*Manis*)

The scaly pangolin (Figure 4.2.23) is a placental mammal from Africa and Asia. It has broad, horny scales rather than hairs on its body. It has no teeth and its sticky tongue is over 30 cm long. The pangolin has small eyes and long, powerful claws. It also boasts a large, heavy tail that measures about the length of the rest of the body.

Head and body length: 30–88 cm (not including tail)

Body weight: 4.5–27 kg

Figure 4.2.23 *Manis* (pangolin)

ISBN 978 0 6557 0026 5

PRACTICAL ACTIVITY 13

Numbat (*Myrmecobius*)

Figure 4.2.24 *Myrmecobius* (Australian marsupial anteater)

The Australian marsupial anteater, or numbat (Figure 4.2.24), has short hair with a characteristic striped pattern. It has small eyes, about 50 small teeth and a long, sticky, cylindrical tongue. The forelimbs end in powerful claws.

Head and body length: 17–28 cm

Body weight: 275–450 g

METHOD

Use the information above to prepare a summary report that focuses on the evolutionary relationships between the different species. Your report should include:

- common features shared by the different species
- differences between the species; include a description of a feature that makes each animal well adapted to its particular lifestyle
- a phylogenetic tree illustrating the evolutionary relationships between the different species; outline the features you used to group as well as distinguish between the different species
- some evidence that suggests divergent evolution from a common ancestor.

Despite belonging to different taxonomic groups, these animals share some characteristics in common. Outline the argument that these animals also represent an example of convergent evolution.

SUMMARY REPORT

ISBN 978 0 6557 0026 5

PRACTICAL ACTIVITY 14

Evidence of evolutionary relationships between selected marsupials

Suggested duration: 60 minutes

INTRODUCTION

In this activity you will consider similarities and differences in morphology and lifestyle for five selected marsupials, as well as examine and compare samples of their genomic data. You will use the information to infer evolutionary relationships between the species.

AIM

- To consider some similarities and differences between selected marsupials.
- To draw conclusions about evolutionary relationships between selected marsupials based on comparative genomics data.

BACKGROUND

Marsupials are a unique and diverse infraclass of mammals represented by more than 140 species, mostly endemic to Australia, with a small number of examples in the broader Australasian region and the Americas. The group diversified and flourished, particularly on the Australian island continent, after it became separated from the supercontinent Gondwana approximately 99 million years ago.

The infraclass Marsupialia is represented by four different orders.

- *Diprotodontia*—this group includes 80 different species, all characterised by the presence of two well-developed incisor teeth in the lower jaw; these animals are primarily herbivorous, although some include insects and small reptiles in their diets.
- *Dasyuromorphia*—this group includes 52 species of carnivorous animals; distinguishing features include well-developed canine teeth.
- *Peramelemorphia*—this group includes eight species of omnivorous animals, that is, they feed on both plant and animal matter, including insects, small birds and mammals, as well as grasses and leaves; a range of tooth types cater for a diverse diet.
- *Notoryctemorphia*—this order is represented by only one species, the marsupial mole, a sand-burrowing, desert-adapted animal that feeds mainly on insect larvae.

1 Consider the visual and written data provided for the five different marsupial mammals.

a Describe characteristics common to mammals.

____

____

____

b Describe a key characteristic common to all marsupials.

____

____

PRACTICAL ACTIVITY 14

Common name: tiger quoll	Scientific name: *Dasyurus maculatus*
	• habitat: eucalypt forests and woodlands • head and body length: 40–75 cm • body weight: 2–3 kg • diet: small birds, mammals, reptiles, insects
Common name: Tasmanian devil	**Scientific name: *Sarcophilus harrisii***
	• habitat: eucalypt forests and woodlands • head and body length: 57–65 cm • body weight: 6–7 kg • diet: small birds, mammals, reptiles, fish, insects, also scavenger
Common name: opossum	**Scientific name: *Didelphis virginiana***
	• habitat: forest • head and body length: 35–95 cm • body weight: 4–6 kg • diet: insects, worms, fruit, nuts, small reptiles, also scavenger
Common name: ringtail possum	**Scientific name: *Pseudocheirus peregrinus***
	• habitat: eucalypt forests • head and body length: 30–35 cm • body weight: 1 kg • diet: eucalypt leaves, flowers, fruit
Common name: eastern barred bandicoot	**Scientific name: *Perameles gunnii***
	• habitat: woodland forests and native grasslands • head and body length: 18–45 cm • body weight: 640 g • diet: beetles, worms, grubs, fungi, berries

 ISBN 978 0 6557 0026 5

2 There are only four orders of marsupials. Identify a key feature that distinguishes between the groups.

3 **a** Decide which of the five species in the visual and written data are most closely related. Create a phylogenetic tree to demonstrate these relationships.

b Justify your response to part **a**.

4 The data table below includes the nucleotide sequence for a section of the *RUNX2* gene in mammals. The gene codes for a protein involved in the regulation of skeletal development. This gene is highly conserved in marsupials.

Examine the data for the five marsupials under consideration in this activity.

D. maculatus	G	C	G	G	C	G	G	C	A	G	C	G	G	C	G	G	C	G	G	C	G	G	C	G	G	C	G	G	C	G	G	C	G	G	C	T	G	C	T	G	C	G	G	C	G	G	C	G	G	C	T	G	C	G	G	C	G	G	C	T	G
S. harrisii	G	C	G	G	C	G	G	C	A	G	C	G	G	C	G	G	C	G	G	C	G	G	C	G	G	C	G	G	C	G	G	C	G	G	C	T	G	C	T	G	C	G	G	C	G	G	C	G	G	C	T	G	C	G	G	C	G	G	C	T	G
D. virginiana	G	C	C	G	C	G	G	C	A	G	C	G	G	C	G	G	C	G	G	C	A	G	C	A	G	C	G	G	C	G	G	C	T	G	C	G	G	C	C	G	C	G	G	C	G	G	C	G	G	C	C	G	C	A	-	-	-	-	-	-	-
P. peregrinus	G	C	C	G	C	G	G	C	A	G	C	A	G	C	G	G	C	A	G	C	G	G	C	G	G	C	G	G	C	G	G	C	G	G	C	A	G	C	G	G	C	G	G	C	G	G	C	G	G	C	A	G	C	G	G	C	A	G	C	C	G
P. gunnii	G	C	G	G	C	G	G	C	A	G	C	G	G	C	G	G	C	G	G	C	G	G	C	T	G	C	T	G	C	T	G	C	T	G	C	T	G	C	T	G	C	G	G	C	G	G	C	G	G	C	T	G	C	G	G	C	G	G	C	G	G

a Identify nucleotide differences present in the nucleotide sequence for the different marsupials. Use a highlighter pen to mark the differences.

b Use the space below to construct a new phylogenetic tree for the five species, based only on the genomic data.

ISBN 978 0 6557 0026 5

c Justify the relationships indicated by this phylogenetic tree.

5 Explain the genomic data for the tiger quoll and the Tasmanian devil, given they are a separate species.

6 Conclude whether the marsupials under consideration represent divergent or convergent evolution.

CONCLUSIONS

7 Infer two conclusions from the genomic data for the five marsupial species.

8 Discriminate between the validity of morphological data and molecular data in terms of establishing evolutionary relationships. Refer to the genomic data to support your answer.

9 Describe other evidence that would add weight to the validity of inferences and conclusions about the evolutionary relationships between the marsupials.

 ISBN 978 0 6557 0026 5

PRACTICAL ACTIVITY 15

Classification and identification • Modelling

Examining primate evolution

Suggested duration: 60 minutes

INTRODUCTION

Over approximately 7 million years of evolution, scientists currently recognise five different hominin genera: *Sahelanthropus, Ardipithecus, Australopithecus, Paranthropus* and *Homo*. Fossils representing these taxonomic groups share particular features that clearly set them apart from earlier primates, including the great apes. The fossil record demonstrates that several species belonging to the genus *Homo* have existed over the last 2.5 million years. Of these, *H. neanderthalensis, H. denisova* and *H. floresiensis* are believed to have co-existed with modern *H. sapiens*. With *H. floresiensis* becoming extinct approximately 12000 years ago, *H. sapiens* are the last remaining species of human on Earth.

AIM

- To examine some similarities and differences between the primate groups.
- To consider some of the major trends in hominin evolution.

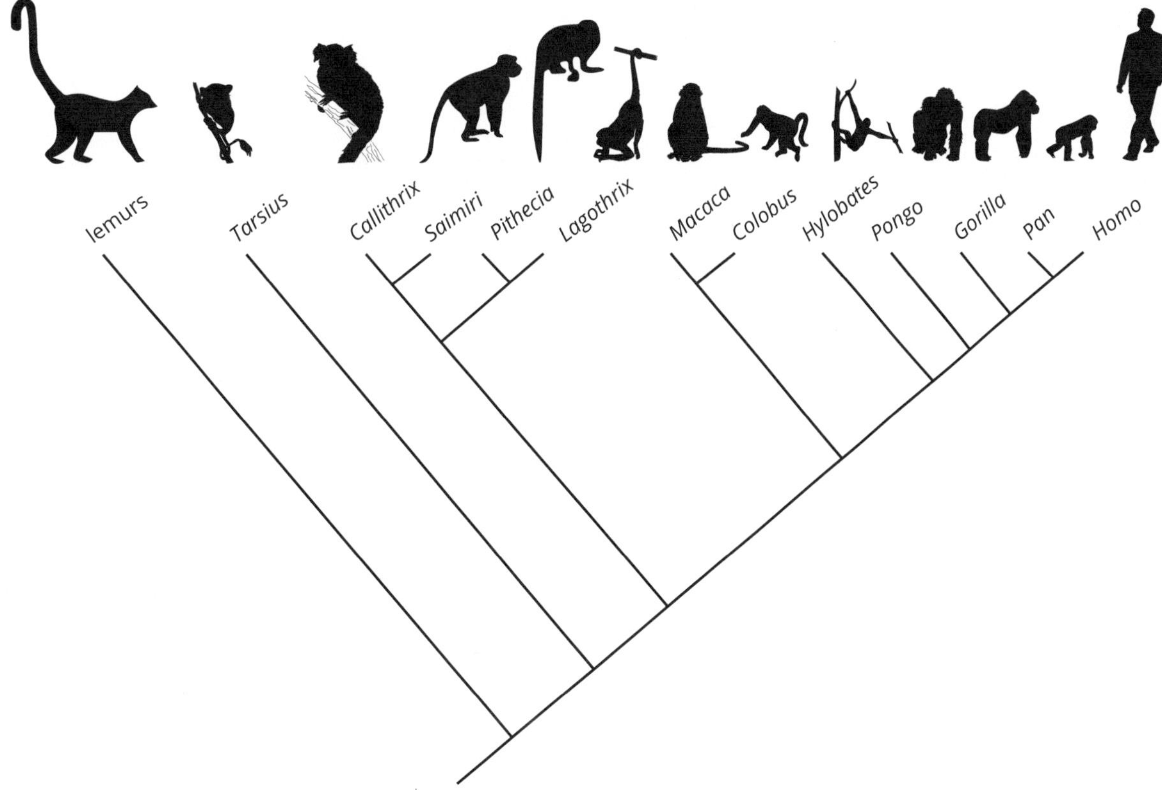

Figure 4.2.25 Phylogenetic tree of Primates

PRACTICAL ACTIVITY 15

1 Humans belong to the order Primates along with lemurs, lorises, tarsiers, old world monkeys, new world monkeys, gibbons and the great apes, which include orangutans, gorillas and chimpanzees. Complete the table below, describing the similarities in the features listed for all of the primates.

Characteristic	Description
hand	
teeth	
eyes	
nose	
brain	

2 Members of the family Hominidae (hominids) include orangutans, gorillas, chimpanzees and humans. Outline the differences between species in the family Hominidae and the rest of the primates.

ISBN 978 0 6557 0026 5

3 Examine the visual information in Figure 4.2.26 contrasting skeletal details between apes and humans. Summarise the key differences between these for the characteristics listed in the table below.

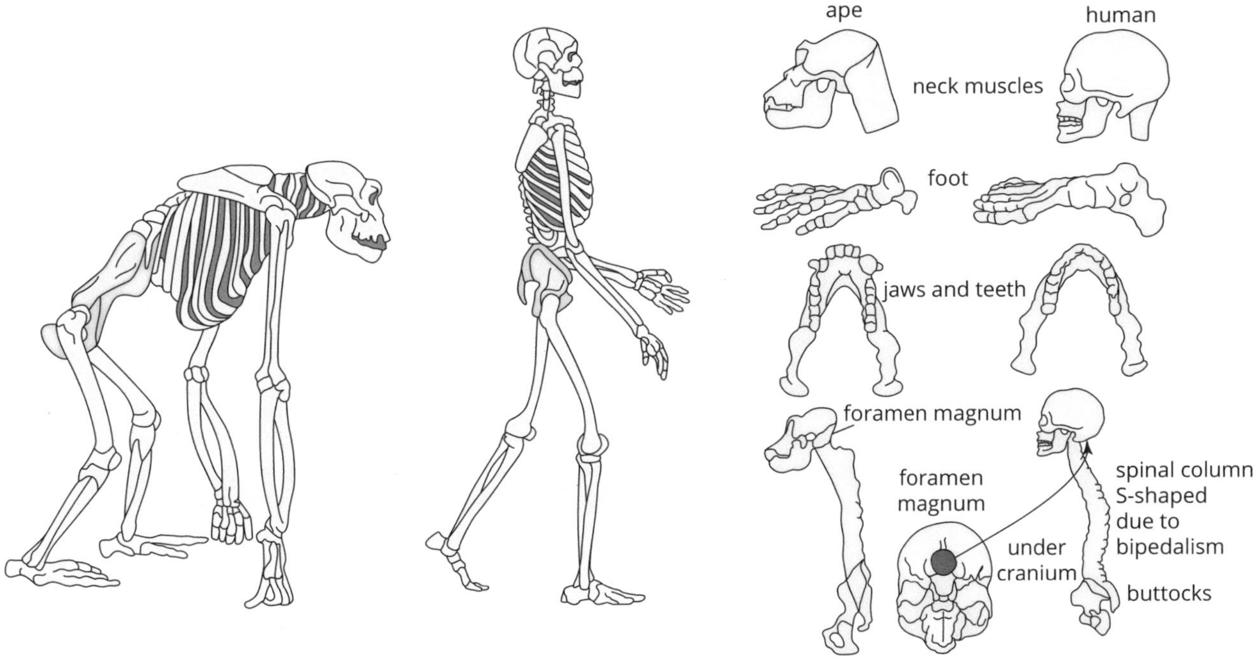

Figure 4.2.26 Skeleton and other skeletal features of humans and apes

Characteristic	Human	Ape
position of foramen magnum		
length of forearms in relation to legs		
feet and toes		
locomotion (bipedal or quadrupedal)		
shape of face		
size of teeth		
shape of jaw		
size of braincase		

ISBN 978 0 6557 0026 5

4 The illustration below shows the skulls of four different hominins, representing different genera of modern and ancestral humans. Use the table below below to summarise the trends observed in skull structure that you observe from skull 1 to skull 4.

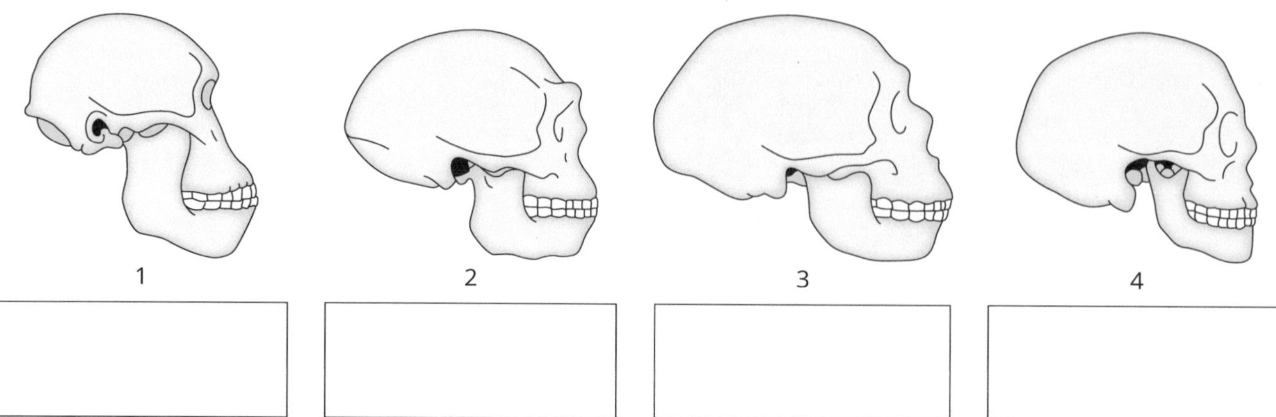

Characteristic	Trend
brow ridge	
shape of face	
teeth size and shape	
jaw size	
relative size of braincase	

5 The skulls represent the following hominin groups: *H. erectus*, *H. sapiens*, *Australopithecus* and *H. neanderthalensis*. Use the box below each skull to correctly assign its name.

6 Outline an advantage of each of the following evolutionary changes to humans.

a bipedalism

b increased brain size and more complex cerebral cortex

 ISBN 978 0 6557 0026 5

PRACTICAL ACTIVITY 15

7 The phylogenetic tree in Figure 4.2.27 illustrates the evolutionary timeline for different hominin species. The chimpanzee is included as an outgroup.

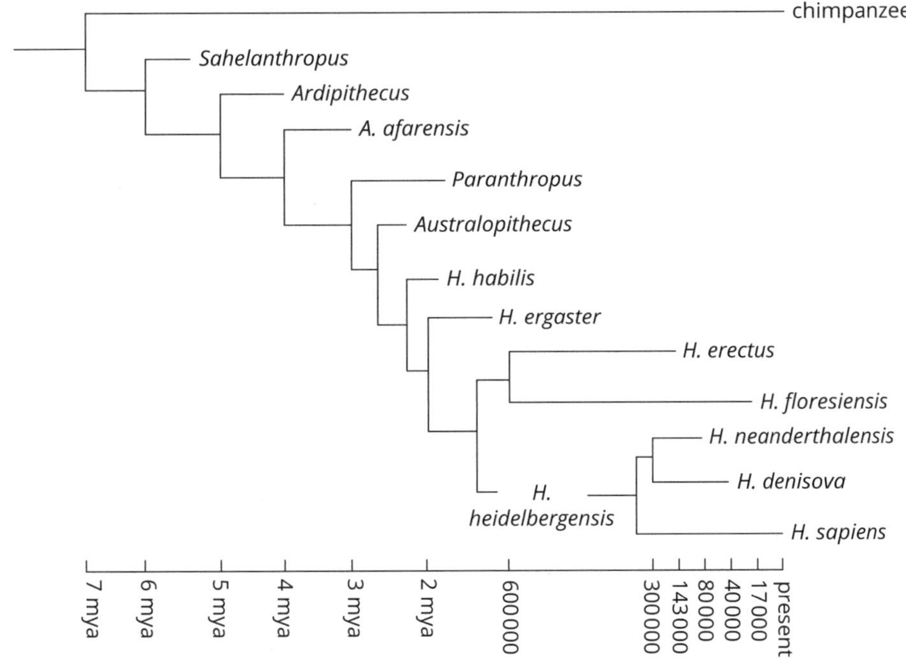

Figure 4.2.27 Evolutionary timeline of hominin species

a According to this model, which of the great apes is the most closely related to hominins?

b Approximately how long ago did the earliest hominin and the species of ape identified in part **a** share a common ancestor?

c List the various hominin genera in order from earliest to most recent.

d How long ago did Neanderthals share a common ancestor with the Denisovans?

e How long ago did modern humans share a common ancestor with Neanderthals?

CONCLUSIONS

8 Outline how to distinguish between hominids and hominins.

9 Describe the key patterns observed in the fossil evidence of hominin evolution.

PRACTICAL ACTIVITY 16

Case study • Modelling

The changing face of human evolution

Suggested duration: 60 minutes

INTRODUCTION

Examination of the evidence related to human evolution provides us with a classic example of the changing nature of evolutionary theory. This truth has never been more evident than at the present time, with recent significant fossil discoveries shedding new light on our unfolding understanding of the evolution of our own species.

AIM

- To evaluate some popular ideas about the evolution of our own human species and culture.

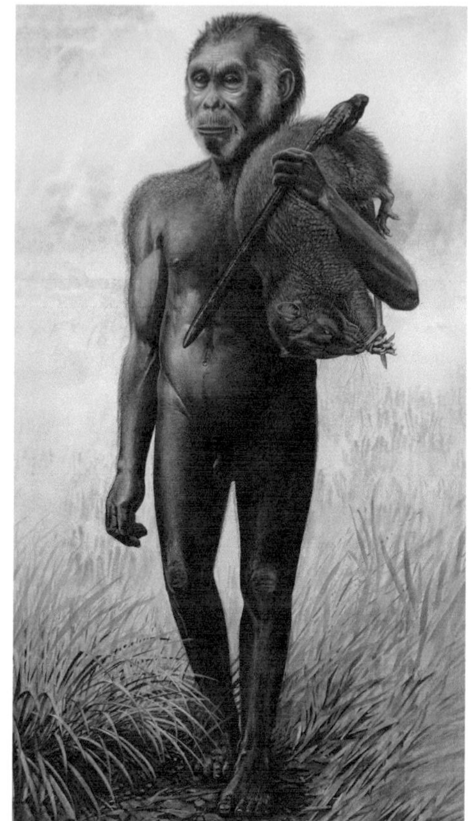

Figure 4.2.28 An artist's impression of reconstructed *H. floresiensis*.

PART A • *HOMO* HOBBIT—THE LOST TRIBE OF FLORES

In 2003, during a research expedition to the Indonesian island of Flores focusing on understanding early human migration from Asia to Australia, a team of archaeologists stumbled upon what appeared to be fossilised human-like remains (Figures 4.2.28 and 4.2.29). Of particular interest was the small stature of the species. The spectacular find was to alter the course of our thinking about human evolution and what it means to be human.

A great deal of evidence was uncovered at the site—anatomical evidence and evidence that provided insights into aspects of the lifestyle of this species. This included:

- a thigh bone indicating the individual had a fully-upright stance and adults stood at about 1 m tall
- a skull revealing a relatively small braincase (380 cm^3), small canine teeth, a flattened nose, protruding mouth and reduced chin
- artefacts including cooking utensils and stone tools
- evidence of fire
- skeletal remains of some animals, including a juvenile stegodon (dwarf elephant)
- bones dated by radiometric analysis found to be around 18 000 years old.

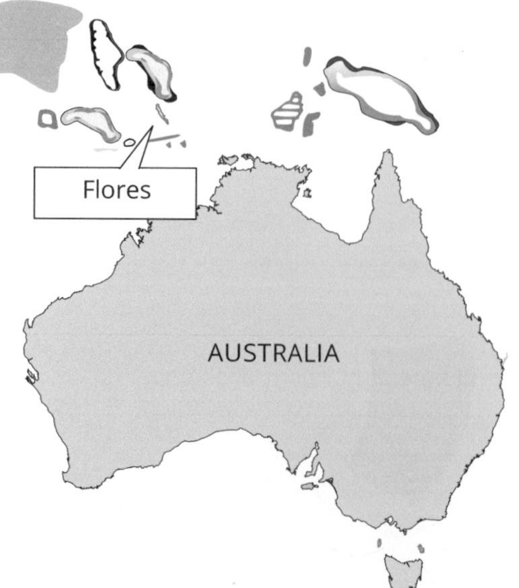

Figure 4.2.29 The island of Flores forms part of the Indonesian archipelago.

 ISBN 978 0 6557 0026 5

PRACTICAL ACTIVITY 16

1 Summarise what the evidence reveals about the potential intelligence and lifestyle of *H. floresiensis*.

2 Our modern understanding of human evolution rests in part on patterns that have been observed between older and more recent species of hominin. How does the discovery of *H. floresiensis* challenge these ideas?

The pieces of the puzzle

Anthropologists trying to make sense of this intriguing discovery are confronted with the following key pieces of information.

- *H. floresiensis*, while much smaller than other species of early human, appears to share a number of features with *H. erectus*.
- *H. erectus* was also a user of tools and fire and is known to have migrated out of Africa to Europe and Asia, including Java.
- The Indonesian island of Flores has existed as an island for over a million years.
- The island of Flores features its own unique species (some extinct) of flora and fauna, including giant rodents, giant lizards and dwarf elephants.

The phylogenetic tree shown in Figure 4.2.30 illustrates a possible pathway of human evolution.

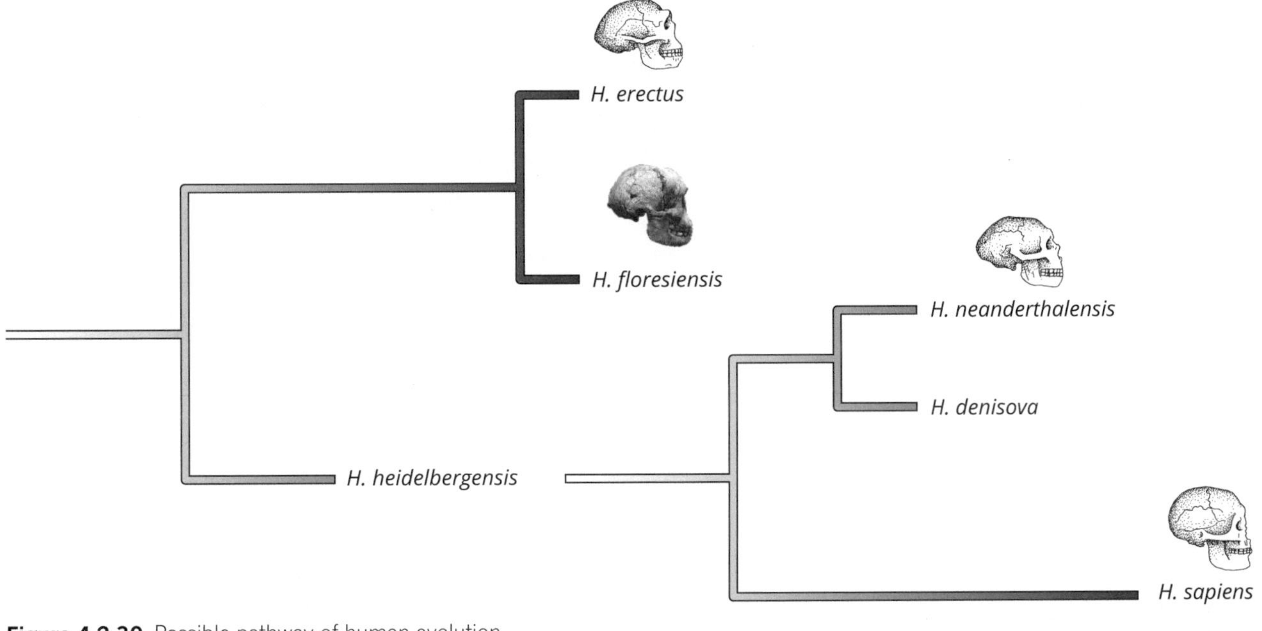

Figure 4.2.30 Possible pathway of human evolution

3 Describe the evolutionary relationship between *H. sapiens* and *H. floresiensis*.

4 Use the phylogenetic information to account for the speciation of *H. floresiensis* on the Indonesian island of Flores.

5 All species of humans are now extinct except for *H. sapiens*. Suggest an explanation to account for the extinction of:

a *H. neanderthalensis*

b *H. heidelbergensis*

PART B • MILESTONES IN THE BIG PICTURE

In 1974 in Hadar, Ethiopia, paleoanthropologist Donald Johanson and his student Tom Gray found the fossilised skeletal remains of what appeared to be an early hominin. The skeleton possessed features such as relatively short legs similar to chimpanzees, but was also bipedal, a feature characteristic of more recent human species. The braincase was small with pronounced canine teeth. The 25-year old female skeleton, dated at approximately 3.2 million years old, was affectionately named Lucy. Lucy was classified as a member of *Australopithecus afarensis*.

In 2003, fossilised remains of the species that would become known as *H. floresiensis* were discovered in a cave on the Indonesian island of Flores. Less than 10 years later, in 2012, another discovery of hominin fossils revealed another new human species in a Denisova cave in Siberia. While only a single finger bone and two molar teeth were recovered for this species, named *H. denisova*, DNA analysis has been able to reconstruct almost the entire genome of the species. Genomic comparisons have placed this hominin alongside *H. neanderthalensis* in terms of a close evolutionary relation. The fossils have been dated at around 80000 years old, revealing they co-existed with *H. neanderthalensis* and *H. sapiens*.

Then in 2013 a spectacular new find uncovered what scientists believe to be the possible burial ground of the oldest human fossil of the genus *Homo*. The previously unknown species was named *Homo naledi* after the fossilised remains of 15 skeletons were found in the Dinaledi chamber of the Rising Star cave system northwest of Johannesburg in South Africa. The members of this newest species of the human family stood at about 1.5 m tall. While features such as curved fingers and a broad ribcage are similar to *Australopithecines*, the relatively shorter arms than legs, non-grasping toes, shape of the pelvis and bones of the feet clearly demonstrate bipedalism. The anatomical features described, as well as the relative size and shape of teeth, jawbone and skull, have resulted in the species being placed into the genus *Homo*. No tools were discovered with the fossils.

Each time these new and spectacular discoveries occur, they add to the big picture of our understanding of our human origins.

6 Examine the skull profiles for the different hominin species shown in Figure 4.2.31.

A. afarensis
- 3.2 mya
- pronounced canine teeth
- 1.5 m tall
- braincase: 500 cm^3
- no evidence of tools
- bipedal

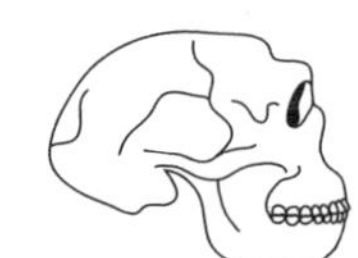

H. erectus
- 1.6 mya–100000 years ago
- braincase: 1000 cm^3
- fossils associated with stone tools, burnt stone and animal bones, charcoal and ash deposits

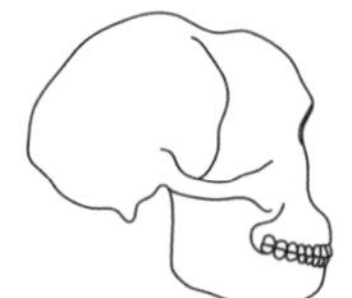

H. floresiensis
- 95000–13000 years ago
- braincase: 380 cm^3
- fossils associated with stone tools and animal remains, cooking utensils

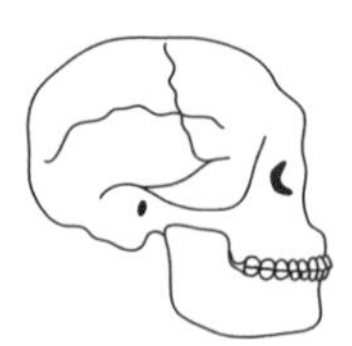

H. sapiens
- 200000 years ago–present
- braincase: 1300 cm^3

Figure 4.2.31 Skull profiles of different hominin species

 ISBN 978 0 6557 0026 5

PRACTICAL ACTIVITY 16

a Compare the *H. naledi* skull in Figure 4.2.32 to the hominin skulls in Figure 4.2.31. Suggest why scientists have classified *H. naledi* in the genus *Homo* rather than *Australopithecus*.

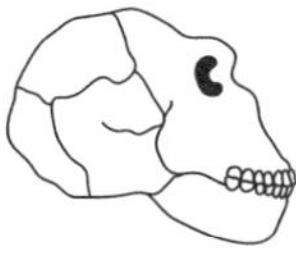

H. naledi
- skull slightly larger than a chimpanzee
- 1.5 m tall
- relatively small teeth with reduced canines
- braincase: small but rounded
- bipedal

Figure 4.2.32

b Where would you place *H. naledi* in this evolutionary order? Explain your reasoning.

c What does the evidence suggest about the relative age of the *H. naledi* fossils? Explain your response.

7 Look carefully at the Rising Star cave system and the location of the fossil chamber (Figure 4.2.33).

Figure 4.2.33 Rising Star cave system showing the discovery chamber for *H. naledi*

Scientists have debated how the skeletal remains arrived in the chamber, which has difficult access via a long, narrow chute. Given the evidence, theories suggesting that the bones were carried in by predators or by underground river systems or flooding have been discounted.

a Suggest how else the skeletal remains may have arrived at this final resting place.

b What does this reveal about *H. naledi*? Why is this surprising?

c Account for the absence of tools found with the fossils.

8 Suggest why *H. naledi* has remained undiscovered for so long.

CONCLUSIONS

9 Describe the evidence of culture in:

a *H. floresiensis*

b *H. neanderthalensis*

c *H. naledi*

10 Outline how our understanding of our uniqueness as a human species is challenged by discoveries of ancient hominin fossils.

FURTHER INVESTIGATION

Use online resources to search for further information about *H. naledi*.

Dating *H. naledi* fossils has revealed some surprising results.

- How long ago is *H. naledi* believed to have lived?
- List some questions this estimated dating raises about the species.
- Explain why this estimated age is significant in terms of our understanding of human evolution.

 ISBN 978 0 6557 0026 5

PRACTICAL ACTIVITY 17

Case study • Modelling

The first Australians—migration and genetic divergence

Suggested duration: 60 minutes

INTRODUCTION

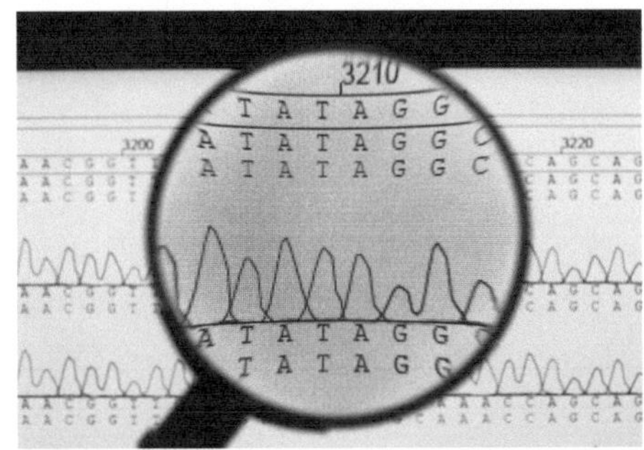

Bioinformatics is the practice of using computer technology to gather, store, analyse and manipulate biological data. It has particular relevance for genetic research, where significant volumes of data are routinely under investigation. Sequencing genomes for entire organisms is an example of how bioinformatics can analyse large quantities of data efficiently. Comparing the DNA sequences of different species and identifying the degree of similarity and difference between them provides scientists with an indication of the evolutionary relationship between the species, and an estimation of how long ago the species diverged. This is also true for different human populations—the degree of similarity and difference in the genomes of people from different populations indicates approximately how long ago they diverged from one another.

Various factors contribute to the evolution of species over time, and the divergence of different kinds of organisms from ancestral species. Change in an organism's DNA (mutations) is responsible for the variation we see in particular features, such as dark- and light-coloured hair. Presuming mutation rates are relatively constant, the rate of change in the DNA of related organisms is useful in estimating how long ago ancestral groups diverged from one another. Environmental conditions also play a key role, because some features make organisms well adapted to particular environments, giving them a survival advantage, while others do not. The migration of populations of a species to different regions so that the populations are isolated from one another also contributes to divergence, particularly when the populations are subjected to different environmental pressures.

The migration of humans to the Australian continent is a case in point. When the first people arrived on what is now the Australian continent, the landscape was very different from what we recognise today. About 50 000 years ago, Earth was in the midst of an ice age, with much of the planet's sea water locked up in the polar ice caps. This resulted in a decrease in sea level, exposing the continental shelf so that land bridges connected the Australian mainland to what is now Tasmania in the south and Papua New Guinea in the north. This much larger landmass is referred to as Sahul. To the northwest, the Indonesian archipelago was joined to Southeast Asia in a landmass called Sunda (Figure 4.2.34 on page 192).

AIM

- To consider genomics data related to recent genetic divergence and migration in human evolution, in particular in relation to Indigenous Australians.

Indigenous ingenuity—an evolutionary success story

According to recent genomics and archeological research reported in the science journal *Nature*, the oldest continuing population of people on Earth is Indigenous Australians. Data in the study compared the sequenced genomes of 83 Indigenous Australians representative of communities across the geographic range of the continent. It also compared the genomes of Indigenous Australians with other groups, including native Papuans and populations from Asia, Europe and Africa. Similarities and differences in the genomes led to the following conclusions.

- Ancestors of Indigenous Australian and Papuan groups migrated in a single exit wave from the African continent approximately 72 000 years ago.
- The ancestral Indigenous Australian/Papuan population reached Sahul around 50 000 years ago.
- Indigenous Australian and Papuan populations were already isolated from one another by around 37 000 years ago.
- Ancestral Indigenous Australian populations migrated throughout the Australian continent.

PRACTICAL ACTIVITY 17

- As early as 31 000 years ago, significant genetic diversity existed between populations scattered on the Australian continent, and this was particularly marked in populations between the east and west.

Other evidence revealed by the study includes the following.

- Some similarities exist in the genomes of Indigenous Australians and the Denisovans (an ancient, extinct human species that inhabited east Asia and Siberia; Denisovan extinction is estimated at approximately 41 000 years ago).
- Some similarities exist in the genomes of Indigenous Australians and Neanderthals (an ancient, extinct human species that inhabited Europe and parts of Asia; Neanderthal extinction is estimated at approximately 40 000 years ago).

1 Use the map below (Figure 4.2.34) to prepare a visual summary of human migration to the Australian continent. Include:
   - labels for the continents Sahul and Sunda
   - arrows to illustrate the migration path from Africa to Australia on the inset map
   - arrows to illustrate the migration path from Sunda to Sahul
   - notes stating how long ago these migrations occurred
   - labels for areas of exposed continental shelf.

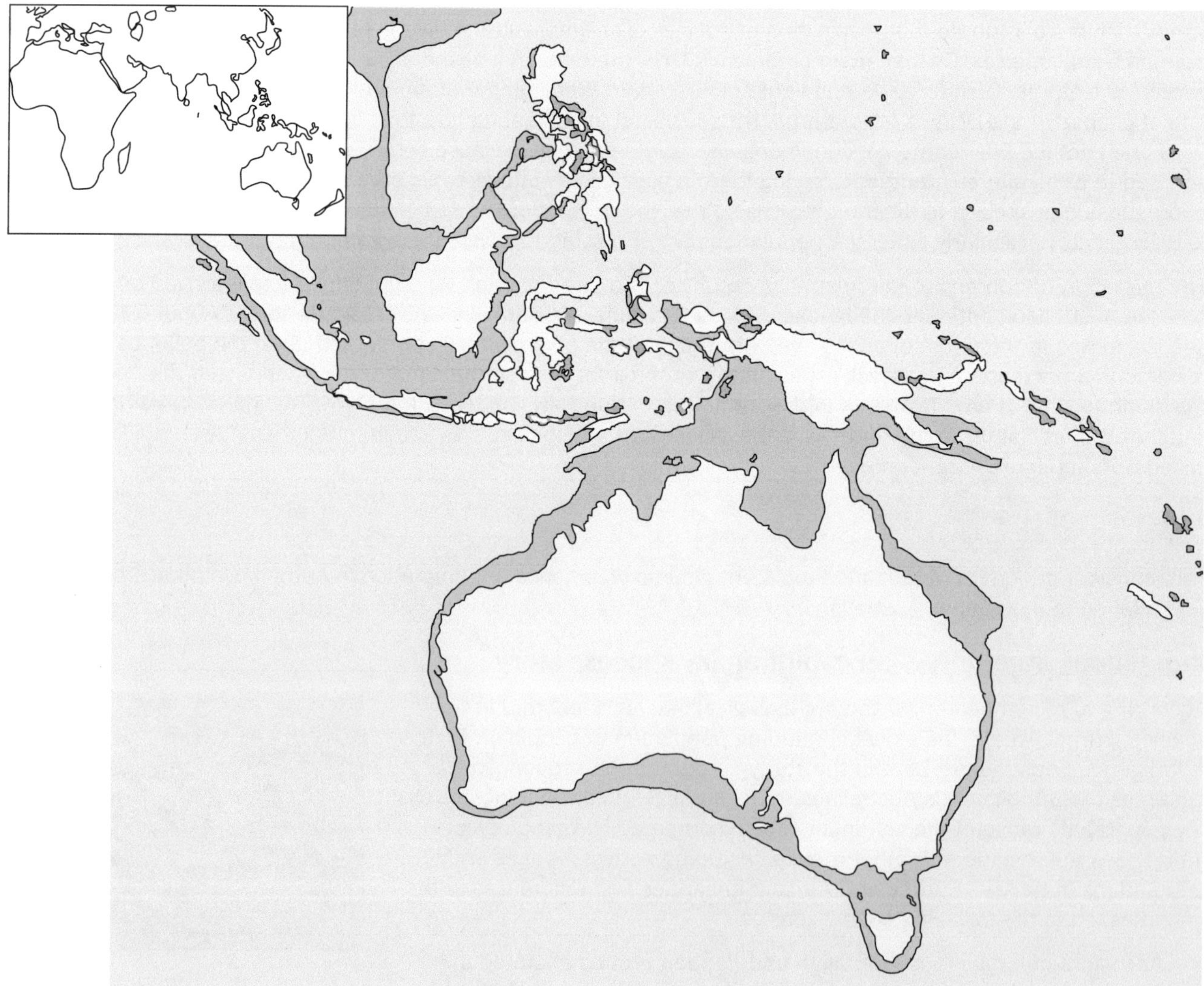

Figure 4.2.34 Present-day Australia with Sahul and Sunda boundaries shaded; inset map of Africa, Eurasia and Australasia

ISBN 978 0 6557 0026 5

2 Account for the similarities in the genomes of Indigenous Australians and the Denisovans and Neanderthals.

3 On what basis have scientists arrived at the following conclusions?

a An ancestral population of Indigenous Australians and Papuans migrated from Africa around 72 000 years ago.

b Indigenous Australians and Papuans diverged approximately 37 000 years ago.

4 The 274 islands that comprise the Torres Strait Islands are located in the waters between the tip of the Cape York peninsula and New Guinea. Explain the likely pathway that led to distinct populations on these islands and mainland Australia.

5 Suggest a reason for the significant genetic diversity between Indigenous Australian populations in the east and west of the Australian continent.

ISBN 978 0 6557 0026 5

EXAM QUESTIONS

Multiple-choice questions

Question 1 VCE Biology 2019 (A) 28

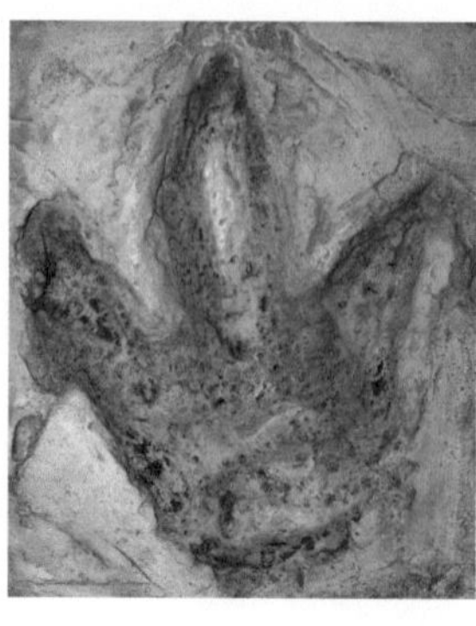

Consider the close-up image of a dinosaur footprint discovered by scientists.

This type of fossil is best described as

A. preserved remains.

B. a petrified fossil.

C. a trace fossil.

D. a cast.

Question 2 VCE Biology 2017 (A) 30

In the 18th century, farmer Robert Bakewell separated large, fine-boned sheep with long, shiny wool from his native stock to interbreed for future sheep flocks.

This is an example of

A. genetic fitness.

B. natural selection.

C. selective breeding.

D. allopatric speciation.

Question 3 VCE Biology 2019 (A) 26

Consider the diagram below showing the gene pool of a population over 20 generations.

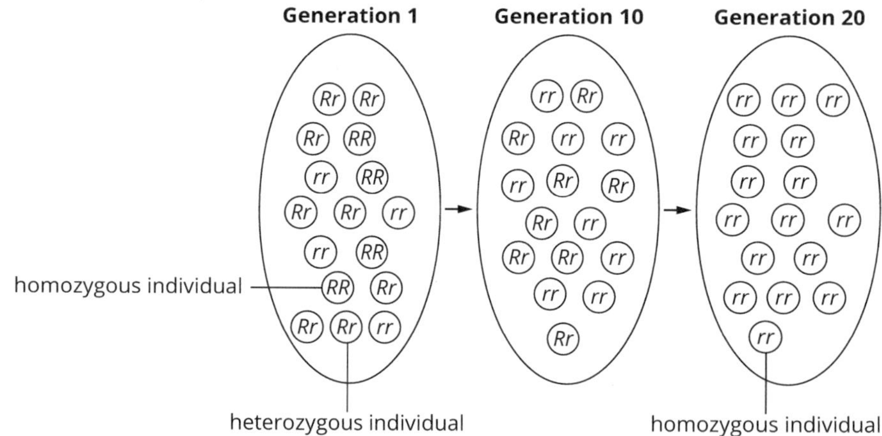

It would be correct to conclude that, over the 20 generations

A. genetic diversity is increasing in this population.

B. individuals with the genotype *RR* had a selective advantage in this population.

C. the frequency of each allele is equal in Generation 1 but not in other generations.

D. new advantageous alleles for this gene were introduced as individuals joined this population.

EXAM QUESTIONS

Question 4 VCE Biology 2017 (A) 32

The phylogenetic tree below represents one model of the order and approximate time of appearance of the major groups of living organisms, and includes four groups represented by the letters R, S, T and U.

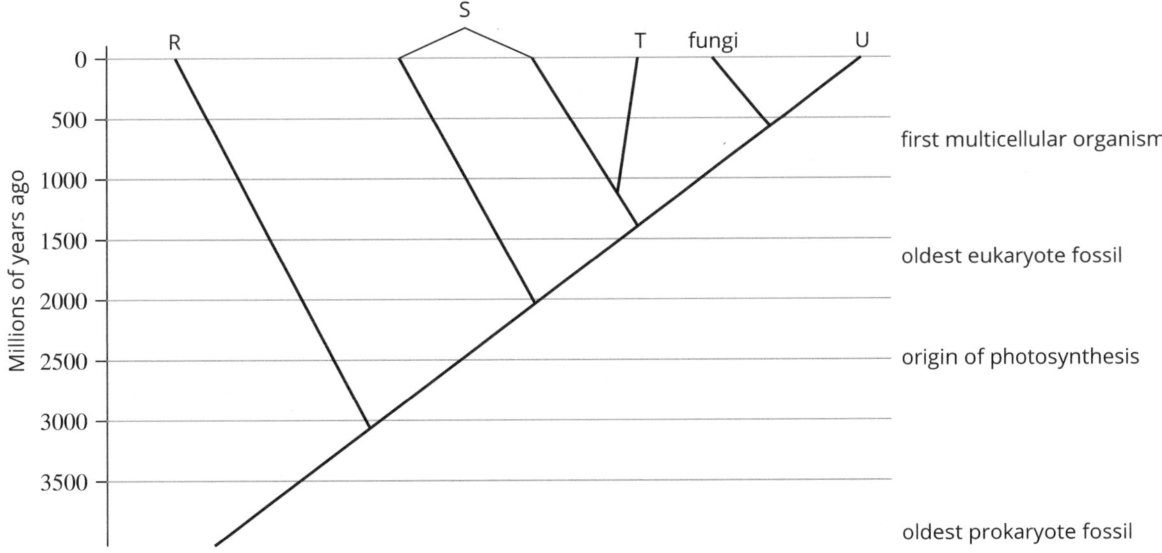

Which of the following shows the correct placement of the organisms on the phylogenetic tree?

A. R—animals, S—plants, T—bacteria, U—protists

B. R—bacteria, S—protists, T—plants, U—animals

C. R—protists, S—animals, T—bacteria, U—plants

D. R—plants, S—animals, T—bacteria, U—protists

Question 5 VCE Biology 2017 (A) 35

Modern African *Homo sapiens* do not contain Neanderthal DNA. Modern non-African *H. sapiens* contain a small percentage of Neanderthal DNA because of interbreeding between Neanderthals and *H. sapiens*. This interbreeding is thought to have occurred within the time period 65 000 to 47 000 years ago. A recent study has found *H. sapiens* DNA in the genomes of 100 000-year-old Neanderthal remains.

From this new discovery, it would be reasonable to conclude that

A. modern Africans are the descendants of Neanderthals.

B. there was an early migration of *H. sapiens* out of Africa before 100 000 years ago.

C. the ancestors of modern Africans migrated from Europe to Africa between 65 000 and 47 000 years ago.

D. approximately 100 000 years ago, Neanderthals bred with *H. sapiens* in Africa before the Neanderthals spread to the rest of the world.

Question 6 VCE Biology 2017 (A) 36

In recent years, scientists have discovered that Neanderthals took care of their elderly relatives, used burial rituals for their dead and gave symbolic meaning to natural objects. Studies have also shown that Neanderthals used complex methods to obtain sharp stone implements and produce glues to attach sharp stones to spears.

These discoveries suggest that Neanderthals

A. had bigger brains than previously thought.

B. had a highly developed culture.

C. lived a solitary lifestyle.

D. used a written language.

EXAM QUESTIONS

Question 7 VCE Biology 2018 (A) 37

Members of the order Primates are mammals.

Which combination of features is common to all primates and distinguishes them from other mammals?

	Feature 1	Feature 2	Feature 3
A.	forward-facing eyes	sloping forehead	fur or hair
B.	binocular vision	opposable thumbs	fully rotating shoulder joints
C.	parabolic jaw	tail	nails instead of claws
D.	even-sized teeth	arms longer than legs	bipedal stance

Question 8 VCE Biology 2016 (A) 38

In India, a group of scientists was studying fossils from a coal deposit formed during the Permian period (290–245 million years ago). They found three fossil species from the same genus in different levels (strata) of the coal. When radiocarbon dating on these fossils was performed, it showed exactly the same levels of carbon-14 in all three fossil species. The data is summarised in the table below.

Fossil species	Depth at which fossil was found in the coal deposit (m)	Proportion of carbon-14 (%)
Gangamopteris major	6.2	0.0001
Gangamopteris obliqua	8.1	0.0001
Gangamopteris clarkeana	4.7	0.0001

Which one of the following is the correct conclusion to draw from these findings?

A. There is no evolutionary relationship between these three fossil species.

B. *G. clarkeana* is the common evolutionary ancestor of *G. major* and *G. obliqua*.

C. As carbon dating is a more reliable dating technique than analysis of strata in coal deposits, the fossils of *G. major*, *G. obliqua* and *G. clarkeana* are all of the same age.

D. An analysis of strata in coal deposits is a more reliable dating technique than carbon dating for Permian fossils; the fossil of *G. major* is younger than the fossil of *G. obliqua*.

Use the following information to answer questions 9 and 10. VCE Biology 2016 (A) 39 and 40

Cytochrome c is a protein that consists of 104 amino acids. Many of these 104 sites on cytochrome c contain exactly the same amino acid across a large range of organisms. There are, however, some differences at certain sites. It is hypothesised that different organisms, all containing cytochrome c proteins, descended from a primitive microbe that lived over 2 billion years ago.

The table below uses the three-letter codes for various amino acids found at specific sites for each organism.

Molecular homology of cytochrome c					
Organism	**Site 1**	**Site 4**	**Site 11**	**Site 15**	**Site 22**
human	Gly	Glu	Ile	Ser	Lys
pig	Gly	Glu	Val	Ala	Lys
dogfish	Gly	Glu	Val	Ala	Asn
chicken	Gly	Glu	Val	Ser	Lys
Drosophila	Gly	Glu	Val	Ala	Ala
yeast	Gly	Lys	Val	Glu	Lys
wheat	Gly	Asp	Lys	Ala	Ala

ISBN 978 0 6557 0026 5

EXAM QUESTIONS

Question 9 VCE Biology 2016 (A) 39

Using only the data for the molecular homology of cytochrome c, which one of the following organisms is most closely related to the dogfish?

A. *Drosophila*

B. chicken

C. human

D. yeast

Question 10 VCE Biology 2016 (A) 40

Using only the data for the molecular homology of cytochrome c, which pair of organisms is most distantly related to wheat?

A. dogfish and *Drosophila*

B. *Drosophila* and yeast

C. *Drosophila* and pig

D. human and yeast

Short-answer questions

Question 1 (6 marks) VCE Biology 2017 (B) 5

The rufous bristlebird (*Dasyornis broadbenti*) is a ground-dwelling songbird. The rufous bristlebird is found in gardens near thick, natural vegetation and builds nests in shrubs close to the ground. The rufous bristlebird feeds on ground-dwelling invertebrates. It is a weak flyer and is slow to go back to areas from which it has been previously eliminated. Two distinct populations of rufous bristlebird exist in Victoria. The distribution of each population is shown on the map of Victoria below. The distance between Population A and Population B is over 200 km.

a. Define the term 'gene flow' and explain whether gene flow is likely to occur between these two populations. 3 marks

b. Both of the rufous bristlebird populations in Victoria are small. 3 marks

Referring to the theory of natural selection, explain why the rufous bristlebird is at risk of extinction.

____

____

____

____

____

____

Question 2 (3 marks) VCE Biology 2017 (B) 7

In 2013, about 1500 fossil bones of a hominin species were found in a cave in South Africa. From these bones, scientists have managed to construct an almost complete skeleton. The fossil bones have some features in common with those of the genus *Australopithecus*; however, they have enough similarities to the genus *Homo* that scientists have classified the fossil skeleton as belonging to a new species, *Homo naledi*.

a. What are **two** features that the fossil skeleton would need to have in order to be classified in the genus *Homo* and not in the genus *Australopithecus*? 2 marks

____

____

Finding out the age of these *H. naledi* fossils has been both difficult and controversial. A group of scientists claims that the age of the fossils is more than 2 million years and suggests that *H. naledi* might be a 'link' between *Australopithecus* and *Homo*. A second group of scientists has calculated the age of the *H. naledi* fossils to be only about 900 000 years and claims that *H. naledi* cannot be the 'link' between *Australopithecus* and *Homo*. The diagram below indicates the time periods for different *Australopithecus* and *Homo* species.

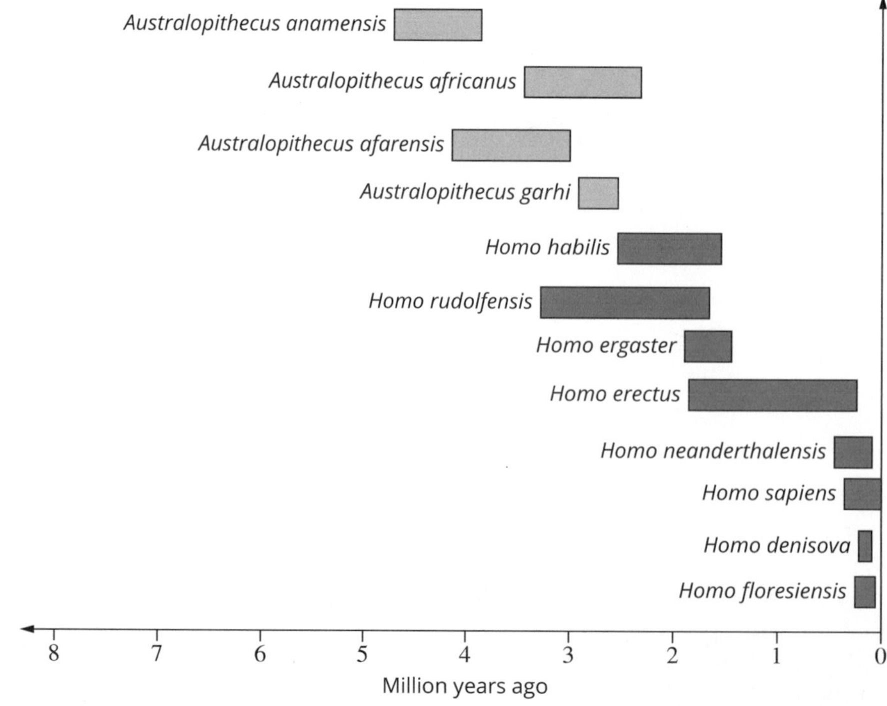

 ISBN 978 0 6557 0026 5

b. If the second group of scientists has correctly dated the *H. naledi* fossils, what evidence from the diagram above supports this group's claim that *H. naledi* cannot be the 'link' between *Australopithecus* and *Homo*? 1 mark

Question 3 (6 marks) VCE Biology 2016 (B) 8

Two species of *Cryptasterina* sea stars are found in coastal Queensland. *Cryptasterina pentagona* is found in warmer water further north, while *Cryptasterina hystera* is found further south in cooler water.

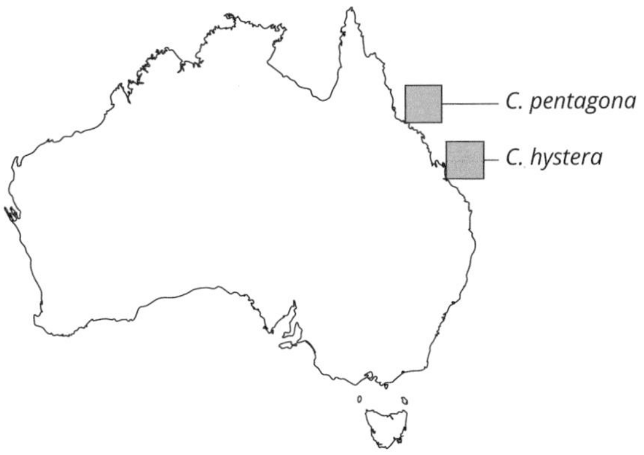

C. pentagona
C. hystera

Researchers have concluded that these two species arose from a recent common ancestor via natural selection. They believe that, over thousands of years, the sea environment has changed, with the boundary line between cold water and warm water moving further north. They have found that water temperature and predation of sea star larvae by cold-water predators are important selection pressures for these sea stars.

a. Using the information **above**, explain how natural selection can lead to differences in phenotypes between these two sea star species. 4 marks

b. One of the phenotypic differences between these two species of sea stars is their method of reproduction. *C. pentagona* reproduces sexually and its sperm and eggs are free-floating in the ocean. *C. hystera* self-fertilises and its fertilised eggs are kept within the sea star until maturity. 2 marks

The researchers found that one species of *Cryptasterina* has a significantly higher diversity of alleles in its gene pool than the other species.

Using this information about reproduction strategies, which species of *Cryptasterina* would you expect to have the highest diversity of alleles? Explain your answer.

ISBN 978 0 6557 0026 5

EXAM QUESTIONS

Question 4 (4 marks) VCE Biology 2013 (B) 10

In humans, severe acute respiratory syndrome (SARS) is a serious form of pneumonia. SARS is caused by a coronavirus that was first identified in 2003. Scientists suspected that the virus had been transmitted to humans from some other animal. Testing was completed on several animal species. Strains of the coronavirus similar to those found in humans were identified in different species of horseshoe bats (genus *Rhinolophus*) and palm civets (*Paguma larvata*).

Samples were taken from the different sources and the virus's RNA from each sample was sequenced.

a. What molecular information would the scientists obtain from sequencing RNA? 1 mark

____

____

The molecular information enabled the scientists to draw an evolutionary tree for different strains of the coronavirus.

The following evolutionary tree was drawn.

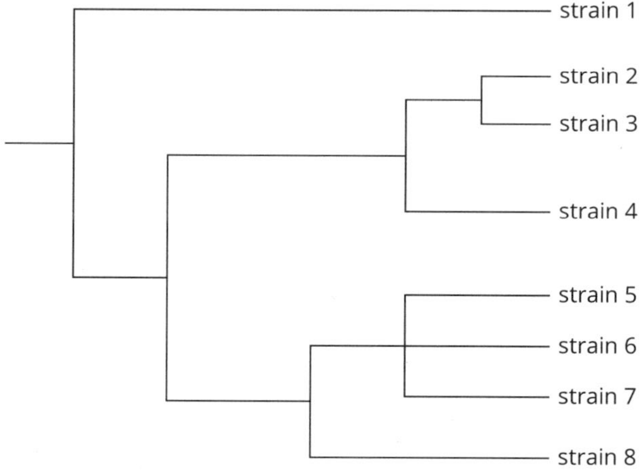

b. Coronavirus strains 2 and 3 are the most similar of the strains. 2 marks

Using your knowledge and the information given in the evolutionary tree, give **two** possible explanations as to why they are the most similar of the strains.

____

____

____

____

Strain 7 is found in palm civets, and strains 5 and 6 in humans. All other strains are found in different species of horseshoe bats.

c. What conclusion can be drawn about the origin of the strain of virus that causes SARS in humans? 1 mark

____

____

 ISBN 978 0 6557 0026 5

EXAM QUESTIONS

Question 5 (7 marks) VCE Biology 2016 (B) 10

Over the past 20 years, a number of new hominin fossils have been discovered. *Homo erectus georgicus* was found near the banks of the Black Sea in Georgia and *Homo naledi* was found in a cave in South Africa.

a. Consider the conditions that may have led to the fossilisation of members of these species. 2 marks

Complete the table below by identifying one condition in the environment of each species that will have made fossilisation possible. The same answer cannot be used for both species.

Species	Environment	Condition
H. erectus georgicus	near the banks of the Black Sea	
H. naledi	cave in South Africa	

Shown below is a photograph of a skull of *H. erectus georgicus*. Scientists compared this skull to that of modern humans (*Homo sapiens sapiens*).

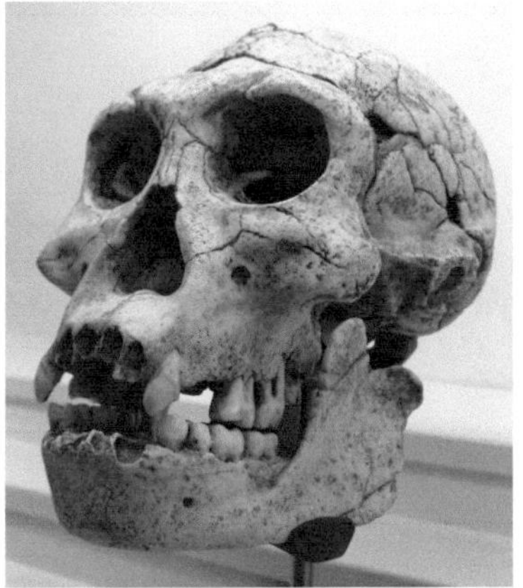

b. Describe any **two** features of the skull shown in the photograph above that allowed scientists to determine that this was a much earlier species of the genus *Homo* than modern humans (*H. sapiens sapiens*). 2 marks

c. Describe **one** structural feature (other than skull structure) of *H. naledi* that would indicate it is a more modern species than members of the genus *Australopithecus*. 1 mark

d. Fifteen different skeletons of *H. naledi* were found in the cave. It was noted that they were all of different ages. 2 marks

Describe **two** pieces of evidence that scientists could have looked for in the cave to indicate cultural evolution within this species.

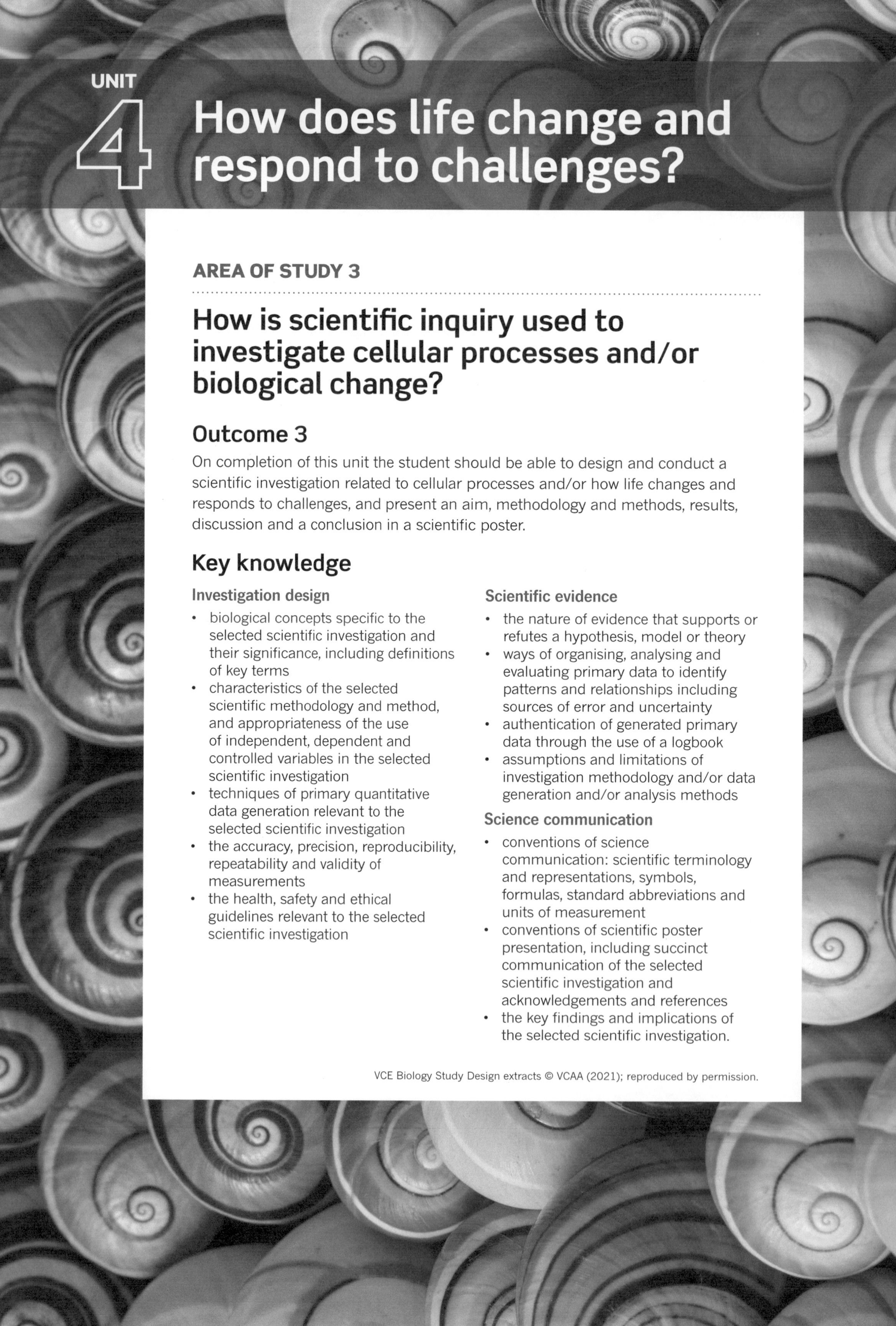

UNIT 4

How does life change and respond to challenges?

AREA OF STUDY 3

How is scientific inquiry used to investigate cellular processes and/or biological change?

Outcome 3

On completion of this unit the student should be able to design and conduct a scientific investigation related to cellular processes and/or how life changes and responds to challenges, and present an aim, methodology and methods, results, discussion and a conclusion in a scientific poster.

Key knowledge

Investigation design

- biological concepts specific to the selected scientific investigation and their significance, including definitions of key terms
- characteristics of the selected scientific methodology and method, and appropriateness of the use of independent, dependent and controlled variables in the selected scientific investigation
- techniques of primary quantitative data generation relevant to the selected scientific investigation
- the accuracy, precision, reproducibility, repeatability and validity of measurements
- the health, safety and ethical guidelines relevant to the selected scientific investigation

Scientific evidence

- the nature of evidence that supports or refutes a hypothesis, model or theory
- ways of organising, analysing and evaluating primary data to identify patterns and relationships including sources of error and uncertainty
- authentication of generated primary data through the use of a logbook
- assumptions and limitations of investigation methodology and/or data generation and/or analysis methods

Science communication

- conventions of science communication: scientific terminology and representations, symbols, formulas, standard abbreviations and units of measurement
- conventions of scientific poster presentation, including succinct communication of the selected scientific investigation and acknowledgements and references
- the key findings and implications of the selected scientific investigation.

Investigating colours of light and the rate of photosynthesis

STUDENT-DESIGNED SCIENTIFIC INVESTIGATION

Students design and conduct a scientific investigation related to cellular processes and/or how life changes and responds to challenges, and present an aim, methodology and methods, results, discussion and a conclusion in a scientific poster.

The investigation draws on knowledge and related key science skills developed across Units 3 and 4 and is undertaken by students in the laboratory and/or in the field.

ASSESSMENT FOR OUTCOME 3

Communication of the design, analysis and findings of a student-designed and student-conducted scientific investigation through a structured scientific poster and logbook entries. The poster should not exceed 600 words.

SUGGESTED DURATION

A minimum of 10 hours of class time should be devoted to undertaking and communicating findings related to Area of Study 3. As per Areas of Study 1 and 2, your teacher will be your primary guide. To assist you, the key steps to follow, relevant poster section and suggested time allocation are set out in the table below.

The sample investigation on page 204 steps you through a controlled experiment, conducted following the designing and planning phase of the investigation. The suggested duration for this part of the investigation includes 20 minutes for set-up and 45–60 minutes for observation and recording. The sample investigation also provides a guide for writing your scientific report in poster format.

USING THIS GUIDE

There is scope for developing an investigation on a range of themes relevant to Units 3 and 4. This investigation is drawn from Unit 3 Area of Study 2 with a focus on the key knowledge and skills related to biochemical pathways, and in particular on photosynthesis.

In this student-designed investigation it is important to carefully consider how to monitor the variable under investigation to achieve valid and reliable results. Safety considerations must also be addressed in the planning process. This guide is intended to take you through approaches to planning, conducting and reporting on scientific investigations.

A risk assessment must be completed prior to beginning the investigation. Relevant social and ethical issues should also be addressed.

Refer to the Toolkit at the beginning of this book for more detailed information on designing and conducting scientific investigations, and presenting a scientific report.

Scientific investigation section	Key step	Relevant poster section/s	Suggested time allocation
Designing and planning your investigation	Step 1: Developing aims, hypotheses and predictions	Title Introduction	60–120 minutes
	Step 2: Determining appropriate methodology and methods	Methodology and methods	60–120 minutes
Conducting your investigation and recording and presenting data	Step 3: Conducting your investigation to generate primary data		120–240 minutes
	Step 4: Recording, organising and presenting your data	Results	120–180 minutes
Discussing your investigation and drawing evidence-based conclusions	Step 5: Analysing and evaluating your data	Discussion Conclusion	60–120 minutes
	Step 6: Referencing	References and acknowledgements	30 minutes
Report on your investigation	Step 7: Preparing your scientific poster	All sections	60–180 minutes

INTRODUCTION

BACKGROUND

Chlorophyll in plants photosynthesises by consuming carbon dioxide, using light energy, to produce glucose and oxygen. There are a few different forms of chlorophyll in plants, possessing their own unique abilities to absorb different wavelengths of light. They all, however, reflect much of the green light.

The question under investigation is: Do green plants photosynthesise at different rates under specific wavelengths of the light spectrum? This question will be investigated by measuring the uptake of carbon dioxide (CO_2) by a green plant under different-coloured light sources.

AIM

To investigate how different wavelengths of light affect the rate of photosynthesis.

HYPOTHESIS

If green plants are exposed to green and red light, then the rate of photosynthesis in a green plant will be greatest under red light.

METHODOLOGY AND METHODS

Remember, your methodology is broader than the methods you have selected for this investigation. It includes a rationale for the approach taken and why this is important to the investigation. The methods are the specific steps taken to generate data during your investigation.

Record a list of the materials required to conduct the investigation. The following is provided as a guide.

MATERIALS

- small sample bottles
- spinach
- 1% sodium hydrogen carbonate ($NaHCO_3$)
- timer/stopwatch or CO_2 sensor
- red and green plastic filters (or cellophane)
- elastic bands
- lamp or light box—compact fluorescent or LED

METHODS

Several factors need to be considered in planning this investigation. These may include:

- identification of variables
- how the experiment will be controlled to ensure accuracy, precision, reproducibility, repeatability and validity of measurements
- the length of time the experiment should be conducted for to achieve meaningful results.

Prepare step-by-step instructions. Remember that these instructions should be clear and easy for someone else to follow. They should be written in the third person.

Remember to consider potential hazards and safety precautions. These should be addressed in your methods, and discussed with your teacher before you proceed.

RESULTS

The investigation results will include quantitative data collected, along with other relevant information. It is critical to use a logbook to authenticate generated primary data and observations, and to record any limitations that become apparent.

Remember to present your data/evidence in appropriate formats to illustrate trends, patterns and/or relationships. Think carefully about how best to present your findings.

DISCUSSION

In the discussion you will interpret and evaluate the analysed primary data collected during the investigation.

The discussion is also where you will evaluate the methodology and methods used. You will identify limitations encountered during the investigation, how the limitations have impacted on the investigation and suggest how to improve, reduce or eliminate these in subsequent investigations.

Results should be cross-referenced to relevant biological concepts. You should also link results to the investigation question and aim, and explain whether or not the data support your hypothesis.

The investigation may prompt suggestions for further investigation, and these should also be outlined in the discussion.

CONCLUSION

The conclusion is a clear and concise statement that summarises the findings of the investigation and responds to the investigation question. In the conclusion you should state whether the hypothesis is supported or refuted, drawing on evidence from the investigation to support this. No new information should be introduced.

REFERENCES AND ACKNOWLEDGEMENTS

Remember to reference and acknowledge all secondary sources used in the course of your investigation. This may include:

- your logbook—raw data and/or previously completed relevant investigations
- worksheets supplied by your teacher
- the school website—support material
- internet sources.

When referencing published sources, follow accepted protocols for presenting bibliographical data, such as APA citation style. For example:

Bruns, E., Armstrong, Z., & Bliss, C. (2021). *Heinemann Biology 2* Sixth Edition. Pearson Australia.

ISBN 978 0 6557 0026 5